THE USES OF SCIENCE IN THE AGE OF NEWTON

PUBLISHED UNDER THE AUSPICES OF THE
WILLIAM ANDREWS CLARK MEMORIAL LIBRARY
UNIVERSITY OF CALIFORNIA, LOS ANGELES

Publications from the
CLARK LIBRARY PROFESSORSHIP, UCLA

1.
England in the Restoration and Early Eighteenth
Century: Essays on Culture and Society
Edited by H. T. Swedenberg, Jr.

2.
Illustrious Evidence
Approaches to English Literature of the
Early Seventeenth Century
Edited, with an Introduction, by Earl Miner

3.
The Compleat Plattmaker
Essays on Chart, Map, and Globe Making in England
in the Seventeenth and Eighteenth Centuries
Edited by Norman J. W. Thrower

4.
English Literature in the Age of Disguise
Edited by Maximillian E. Novak

5.
Culture and Politics
From Puritanism to the Enlightenment
Edited by Perez Zagorin

6.
The Stage and the Page
London's "Whole Show" in the
Eighteenth-Century Theatre
Edited by Geo. Winchester Stone, Jr.

7.
England's Rise to Greatness, 1660–1763
Edited by Stephen B. Baxter

8.
The Uses of Science in the Age of Newton
Edited by John G. Burke

The Uses of Science in the Age of Newton

Edited by
JOHN G. BURKE

Clark Library Professor, 1978–1979

UNIVERSITY OF CALIFORNIA PRESS

BERKELEY ● LOS ANGELES ● LONDON

University of California Press
Berkeley and Los Angeles, California

University of California Press, Ltd.
London, England

Library of Congress Cataloging in Publication Data

Main entry under title:

The Uses of science in the age of Newton.

   (Publications from the Clark Library professorship,
UCLA; 8)
   Includes bibliographical references.
   1. Science—Great Britain—History. 2. Science—Philosophy—History.
I. Burke, John G. II. Series.
Q127.G4U83 1983   509.42   83-1223
                ISBN 0-520-04970-5

Printed in the United States of America

        1 2 3 4 5 6 7 8 9

CONTENTS

v

FOREWORD

During the academic year 1978–79 I had the honor of serving as Clark Professor at the William Andrews Clark Memorial Library of the University of California, Los Angeles. I wish to thank Chancellor Charles E. Young for appointing me to this distinguished position, and Robert Vosper, director of the Clark Library, and the Clark Library Committee for their recommendations on my behalf.

The library was established by William Andrews Clark, Jr., and bequeathed in 1934 to UCLA as a memorial to his father, Senator William A. Clark. The handsome Italian Renaissance-style building housing the collections was designed by Robert D. Farquhar and completed in 1926. Murals and ceiling paintings by Allyn Cox decorate the interior, to which seventeenth- and eighteenth-century furniture lends an atmosphere of elegance and comfort. The library collections are principally representative of seventeenth- and eighteenth-century English culture, certain aspects of nineteenth-century English literature, and fine printing of the nineteenth and twentieth centuries. When it became the charge of UCLA, the collections numbered about eighteen thousand books and manuscripts; at present there are some seventy thousand volumes and over five thousand manuscripts.

The Clark Library staff, headed by Librarian Thomas Wright, could not be more gracious and helpful to scholars who visit and study on the premises. I am most grateful for the cheerful welcome and the continuing assistance they gave me throughout my tenure.

Scholarly lectures presented at the Clark Library during my term as professor became the essays that comprise this volume. In retrospect, the most satisfying occasions were the lively and intellectually stimulating discussions between the seminar participants and the invited lecturers following their presentations. It was at these times that issues of fact and interpretation were raised and thrashed out—sometimes thoroughly and sometimes not, due to time limitations; so I also wish to thank those faculty members, students, and interested lay people, who participated regularly in the seminars, for much of the pleasure and profit I derived from this memorable year.

John G. Burke

CONTRIBUTORS

A. Rupert Hall is Emeritus Professor of the History of Science and Technology at Imperial College, London. He was a chief editor of the five-volume *Oxford History of Technology* (1954–1958). Among his publications are *The Scientific Revolution, 1500–1800* (1954); *From Galileo to Newton* (1963); and *Philosophers at War* (1980). He edited the *Correspondence of Isaac Newton* and is a founding editor of the journal *History of Technology*.

Marie Boas Hall is Emeritus Reader in the History of Science and Technology at Imperial College, London. She is the author of *Robert Boyle and Seventeenth-Century Chemistry* (1958); *The Scientific Renaissance, 1450–1630* (1962); and *Robert Boyle on Natural Philosophy* (1965). She is coauthor of the *Unpublished Scientific Papers of Isaac Newton* (1962) and coeditor of the thirteen-volume *Correspondence of Henry Oldenburg*. Currently, she is writing a history of the Royal Society of London in the nineteenth century.

Earl Miner is Professor of English and Comparative Literature at Princeton University. Among his published works on English literature are *Dryden's Poetry* (1967); *The Metaphysical Mode from Donne to Cowley* (1969); *The Cavalier Mode from Jonson to Cotton* (1971); and *The Restoration Mode from Milton to Dryden* (1974). His writings on Japanese literature include

The Japanese Tradition in British and American Literature (1958); *Japanese Poetic Diaries* (1969); and *Japanese Linked Poetry* (1979).

Richard Olson is Professor of History and Willard J. Keith Fellow in the Humanities at Harvey Mudd College, Claremont, California. He is the author of *Scottish Philosophy and British Physics, 1750–1880* (1975) and *Science Deified and Science Defied* (1982). He edited *Science as Metaphor* (1971).

Albert Van Helden is Professor of History at Rice University. He has published numerous articles on the development of the telescope. He is the author of *The Invention of the Telescope* (1977) and *Cosmic Dimensions from Aristarchus to Halley* (forthcoming) and coauthor of *Divini and Campani* (1981).

Commander David W. Waters is Emeritus Deputy Director of the National Maritime Museum at Greenwich. His publications include *The Art of Navigation in England in Elizabethan and Early Stuart Times* (1958); *The Rutters of the Sea: The Sailing Directions of Pierre Garcia* (1967); and *Science and the Techniques of Navigation in the Renaissance* (1980).

Richard S. Westfall is Distinguished Professor of History at Indiana University. He has published *Science and Religion in Seventeenth-Century England* (1958); *Steps in the Scientific Tradition* (1968); *The Construction of Modern Science* (1972); *Force in Newton's Physics* (1972); and *Never at Rest: A Biography of Isaac Newton* (1980).

INTRODUCTION

In the past decade or more, several sociologists of science have argued that the phrase *uses of science* should no longer be conceived of in terms of the "mere" use or application of scientific knowledge but should be defined in a different way. Steven Shapin, one of the scholars associated with this movement, recently wrote an informative and provocative article entitled "Social Uses of Science," in which he described and evaluated the different approaches employed by historians in studying the development of science in the late seventeenth and eighteenth centuries.[1] Most historians of science, Shapin believes, are disinterested in the social uses of science, neglect them, and even avoid incorporating them into their accounts, because they view the uses of science as occurring apart from and subsequent to the production and evaluation of scientific knowledge. These historians, he writes, assume that

individuals in an esoteric sub-culture generate scientific knowledge by contemplating nature and "rationally" assessing their findings. The context wherein science is produced and judged is argued (or more commonly assumed) to be separable from other contexts.[2]

Such a demarcation, Shapin maintains, gives an inadequate or even faulty account of the development of science. As evidence for his position he describes in some detail the studies of Margaret Jacob, James Jacob, and others, who have produced a body of work (termed by Shapin the "new contextualist" tradition) which argues that science in England during the

Civil War, Interregnum, and Restoration was "powerfully shaped" by social uses. Margaret Jacob, Shapin writes, demonstrates that "conceptions of nature are *tools*, instruments which historical actors in contingent settings pick up and deploy in order to further a variety of interests social as well as technical."[3] James Jacob, in Shapin's view, has shown that Robert Boyle in the 1650s and 1660s "was overwhelmingly concerned to devise a cosmology which might serve to secure a 'moderate,' spiritually governed social order against perceived threats arising from radical sectaries, Hobbists, and other types of philosophical and religious heretics inimical to latitudinarian interests."[4] Further on, Shapin adds:

Indeed, the Jacobs have recently gone so far as to argue that Newtonianism provided "the metaphysical foundations of the Whig constitution" and have given further documentation of the identification between those groups which espoused and disseminated Newtonian natural philosophy and those whose social interests lay in defending and supporting Whig political order and the authority of Low Church Anglican clerics.[5]

These contextualists argue, then, that Newtonian natural philosophy was in large part conceived to legitimate certain political goals and to justify particular religious positions, and was supported and defended by those segments of society sharing common social goals with the theorists. These *social uses* or ends were part and parcel of the intellectual baggage of the scientists involved, conscious and ever present in their minds, and were inextricably involved with technical considerations in the formulation and development of theory. There is no claim that these social uses *determined* the development of natural philosophy. In fact, Shapin writes that the contextualists "do not posit social uses as the sole explanation of scientific change or stability" but that "they also identify a role for the technical interests which we are accustomed to call 'properly scientific.' "[6]

In his article, however, Shapin goes beyond consideration of the social uses of science by asserting that "the technical and practical uses of scientific knowledge has traditionally been as neglected as its social uses. It is as if the practical context in which science has been deployed is an historiographical blindspot."[7] Shapin's meaning becomes clear, I believe, if we consider an earlier reference he makes to the work of Boris Hessen and Robert Merton (the old contextualists?).

What has not been widely accepted is the necessity of referring scientific thought to its social and political context. Nothing is quite as handy for the wholesale dismissal of a disliked mode of practice than the availability of an exemplar which may be pointed to at need as particularly crass and sterile. In this connection Boris Hessen's "Social and economic roots of Newton's *Principia*" has performed yeoman service as strawman, although whether rejection by *fiat* counts as refutation is another matter. And even Merton's cautious sociological approach to scientific "foci of interest" has scarcely been confronted, although perhaps he brought that fate upon himself by his reluctance to extend his contextual techniques to the contents of scientific knowledge.[8]

Hessen's 1931 assessment of the practical context of seventeenth-century science, Shapin continues, was flawed by his individualism, his positivism, and his determinism, but the newer contextualism "cannot fairly be tarred with the same brush used for Hessen." Robert K. Merton's more sensitive study in the same decade was, to Shapin, marred by a lack of boldness and comprehensiveness. No historian (or sociologist), it seems, has followed their lead and repaired the defects in their accounts.

Some of the following essays directly address the theses of the old and new contextualists. Others, in exploring facets of seventeenth- and early eighteenth-century science and technology, raise issues relevant to any discussion of contextualist historiography. All, I believe, contribute to an evaluation of the relative merits of considering *use* in a sociological sense, or of thinking of the uses of science in the traditional way in terms of "mere" use or application.

The first paper, by Earl Miner, permits us to step back and view seventeenth-century science from the perspective of some of England's major contemporary poets from Donne to Dryden. Although natural philosophy was thought to be knowledge of an important kind, it was considered inferior to religious wisdom. In the early seventeenth century millennarian beliefs influenced poetic attitudes toward the new Copernican astronomy; thus, some poets were uncertain whether arguments about its validity mattered, believing that a harmonious world order guided by divine providence was what was really important. The poets' awareness of scientific advance was also affected by their inability to understand the mathematical formulations of the new astronomy and physics, and they were also confused by attempts to reform Epicurean atomism into a Christian mold. There was a gradual adjustment to science, however, and poets *used* it either as symbol or exemplar. Thus, to the troubled John Donne, Copernican astronomy and the telescopic discoveries of Galileo represented the dissolution of order and harmony and the downfall of the hierarchical moral and social order of his age, and his concerns were later echoed by Robert Burton. After the Restoration, science became, in the hands of Samuel Butler, a symbol of "mystick" and useless learning, and in "The Virtuoso," Thomas Shadwell maliciously parodied Robert Hooke's microscopical studies and Martin Lister's systematic natural history observations. To John Dryden, however, the fruits of contemporary science indicated "a new nature has been revealed," and science became an exemplar of his belief in the progress of human knowledge.

In his assessment of the poets' perspective on science, Miner makes two points of some significance for our central theme. In their attacks on science, Miner believes Butler and Shadwell considered the scientists' search for knowledge only from the point of view of its utility and overlooked the fact that knowledge is above all "valuable for intrinsic reasons, for its own sake." Many seventeenth-century scientists sought knowledge without thought of its usefulness, either social or practical, and in time some of this knowledge served to validate Newtonian science. Also, Miner writes, neither the poets nor the scientists of the seventeenth century were able to make the distinction that we now make between the provinces of literature and science. From that time forward, however, scientists attempted, in Miner's words, "to shake off poetics and metaphor, to secularize, and above all to specialize even more narrowly." Contextualists may derive some comfort from this; most scientific works of the time were still written as literature, intended to teach about natural phenomena and where possible to give evidence of divine guidance. A good example comes toward the end of Newton's long disquisition on comets in book three of the *Principia*, where he postulates that the exhalations and vapors from comets replenish the earth's fluids and thereby sustain life.[9] Such references to a providential nature, however, seem to be overwhelmed by Newton's detailed mathematical derivation of a comet's elliptical orbit and by the space devoted to cometary observations. We cannot be certain which motivation predominated, but it appears that it was the technical. In passing, and as a footnote to Earl Miner's essay, it is ironic that Dryden, a champion of science who voiced his praise of the Royal Society's endeavors in eloquent terms, was replaced as poet laureate after the Glorious Revolution by Thomas Shadwell, who pilloried the efforts of scientists in no uncertain terms. Granted that Dryden was a Tory and a convert to Catholicism and Shadwell a true-blue Protestant, their opposite attitudes toward the pursuit of science seems to have made no difference in the affair.

Marie Boas Hall's contribution treats one important facet of the early activities of the Royal Society, the histories of trades; she describes the avowed purposes of such accounts, their scope and content, their reception, and their eventual fate. Compiling histories of trades was a thoroughly Baconian enterprise, but as Marie Hall points out, the Society's Fellows had quite various perceptions of the purpose of the histories. To some there was a promise of immediate utility in the activity—of material benefit to society, for example, by improvements in agricultural practices, or of advantage to the navy by the development of navigational aids and by improved ship design. To others, however, the histories of trades comprised a method that would assist in the acquisition of knowledge of nature—technology in the service of science. There was, of course, another Baconian path—experimentation—to force nature to yield up its secrets. Henry Oldenburg, the first secretary of the Society and the founding editor of the *Philosophical Transactions*, is the central figure in Marie Hall's essay. Oldenburg was aware of the differing approaches, through his friendship with Hartlib and Boyle, and included histories of trades in the *Transactions* since he considered it the goal of the Society to develop "a universal history of nature." To achieve this objective Oldenburg also solicited and published, over the objections of some of the members, accounts by foreign correspondents, one of which, a description of Huygens's spring-balance watch, precipitated his bitter quarrel with Hooke. After 1670 interest in histories of trades began to lag, and fewer appeared in the *Transactions*. Marie Hall suspects that not only Oldenburg but also the members of the Royal Society became gradually disenchanted with the profitability of the enterprise in terms of its contribution to the acquisition of natural knowledge.

Marie Hall's paper reveals an important problem for the historian who describes and analyzes seventeenth-century science. Technology has in numerous instances posed problems that, after scientific study, led to the establishment of scientific principles. As a case in point she cites the example of the steam engine. But attempts to derive natural laws directly from technological practices have not proved fruitful. Still, in the seventeenth century, this approach was considered a part of natural philosophy, along with experimentation, observations of flora and fauna, mathematical studies, and speculations about the inner workings of nature sometimes based on experiment and sometimes the product of pure imagination. Seventeenth-century natural philosophers did not, as we do, delineate various disciplines or separate the various approaches into neat categories, but this is not to say that they were unaware of differences. Oldenburg certainly made a distinction, as Marie Hall shows, between "philosophy" and "useful arts and practices," and others also made demarcations. In 1672, for example, Jacques Rohault described in his *Traité de Physique* three methods of scientific investigation: simple observation; experiments conducted without presuming what might occur; and experiments made as a result of reasoning in order to determine whether or not the reasoning was correct.[10] Thus, to insist that seventeenth-century science was generated and validated in a context involving social uses (along with some technical interests) does not help the historian provide a coherent account of a multifaceted enterprise. Clearer definitions not only of ends but also of the processes (methods) and the products (mathematical laws and theories) of scientific thought and action would appear to be necessary.[11]

Oldenburg, Marie Hall writes, was enormously interested in the development of new scientific instruments, and this facet of science is the subject of Albert Van Helden's article.

Instrumentation contributed to the development of science in several ways. First, there were new observational instruments, the telescope and the microscope, which made visible the satellites of Jupiter, on the one hand, and the gross anatomy of the flea, on the other—resulting in the creation of new areas for scientific investigation. Second, measuring instruments— accurate clocks and telescopic sights, for example—increased the precision of astronomical observations by orders of magnitude. The demands of mathematical and experimental science not only stimulated this revolution in precision instruments based on scientific principles—instruments such as the pendulum clock and the barometer—but also led to the development of a third category, experimental instruments. The vacuum pump, which permitted scientific investigation in an artificial environment, was in the seventeenth century the prime example of the experimental instrument. Scientists not only designed but also made many instruments because there was a lack of skilled mechanics until the closing decades of the seventeenth century. By 1700, scientific investigation without instruments had become inconceivable in many areas of science, and the idea that better instruments would be available as time went on became entrenched.

As Van Helden points out, national prestige was a definite factor in the construction of ever more powerful telescopes— certainly a social use in Steven Shapin's sense. However, in the short run, the major payoff appears to have been the acquisition of new or refined scientific knowledge, such as Newton's discovery of the light spectrum, or Roemer's determination of the velocity of light. The internal development of science received tremendous stimulus, a result that is difficult to reconcile with the contextualist arguments.

In his essay, Richard Westfall focuses on two concerns: the sources of scientific technology in the seventeenth century and the origins of modern science. To study these problems he takes a careful look at the work of Robert Hooke, conceding that the example of one man, while providing evidence, does not settle major issues. Hooke was both a major scientist of the period and a prolific inventor of mechanical contrivances. He was a confirmed Baconian, believing that the purpose of natural philosophy was to satisfy human needs and desires;

but he also worried at times that concern for immediate usefulness might damage the progress of science. Westfall describes three instances when Hooke did attempt to apply scientific principles to technological uses: in the design of a lamp; in determining the best way to trim a sail; and in investigating the dynamics of pendulums and vibrating springs. In each case Hooke was unsuccessful, and Westfall shows that his failures were due to the prevailing level of scientific knowledge. Thus, Hooke's contributions to technology did not stem from the applications of science but were instead the result of his mechanical skill, his fertile imagination, and the practical tradition that he shared with some of his contemporaries. According to Westfall, scientific technology was translated into fact in the seventeenth century only in the area of scientific instrumentation, where technically skillful scientists concentrated on the needs of science itself. Westfall does not contend that the case of Hooke refutes Boris Hessen's general argument that the demands of emergent capitalism presented technological problems that directed the efforts of scientists in the area of physics, but he does find it hard to reconcile the example of Hooke with Hessen's position. Westfall believes that the Scientific Revolution, instead of being a response to external stimuli, was an internal matter, one that involved a "reconstruction of the basic categories of natural philosophy" and an "insistence on the quantitative character of nature." When a sufficient store of scientific knowledge had been amassed, then scientific technology could become a reality; but this did not occur in the seventeenth century.

Westfall is named by Steven Shapin as one of the historians in the "post-Koyrean intellectualist tradition,"[12] who maintain that the social uses of science occur posterior to the production and assessment of scientific knowledge. In this essay, we see that Westfall takes the same position with respect to the practical uses of science. His account of Hooke's endeavors will have to be confronted by anyone who may approach the practical uses of science from the contextualist viewpoint.

The subject of scientific technology arises again in A. Rupert Hall's essay. Hall considers scientific approaches to gunnery and ballistics in the seventeenth century with particular reference to the activities of the members of the Royal Society.

He confronts the argument of Hessen that technical demands determined scientific problem choice and the more qualified assertion of Robert K. Merton that "military needs tended to direct scientific interest into certain fields." At the beginning of a richly detailed description of both mathematical and experimental efforts in the seventeenth century to determine the actual trajectory of an artillery projectile, Hall makes two significant points: first, that contemporary military art would have been incapable of applying mathematical theory; and second, that scientific interest in projectile motion considerably predated the use of guns and gunpowder. Interest in projectile trajectory in the seventeenth century, Hall shows, was transient rather than sustained. Experimental approaches encountered such difficulties as the problem of air-resistance and the effect of a gun's recoil. Mathematicians found the problems of ballistics fascinating, but the best recognized that they were beyond the scope of their analytical abilities.

Hall draws three conclusions from his study. First, he agrees that many seventeenth-century scientists wished to demonstrate the utility of science as applied to artillery problems, but he insists that one can clearly perceive a distinction between those bent on practical application and the "mathematical idealists," who found the trajectory problem intriguing as an intellectual challenge. Second, he believes that claims that problems of mining, navigation, and war commanded the attention of the Royal Society and its members in its early years are exaggerated, though he agrees with Merton that practical questions did have some influence on scientific problem choice—in the case of gunnery, however, yielding precious few results. His final conclusion, similar to that of Westfall, is that scientific investigations that mastered technological problems came only in the following centuries.

The first major practical triumph of science was the perfection in the mid-eighteenth century of methods of determining longitude accurately. Both of the methods that were ultimately successful had been suggested in the early sixteenth century: the measurement of lunar distances was proposed by Johann Werner in 1514 and the transport and comparison of accurate chronometers by Gemma Frisius in 1530. The problem was so well known that, as Earl Miner mentions, poets were aware of it; John Dryden hoped that its solution would be provided by the progress of science. To seventeenth-century scientists, the solution of the longitude problem, A. R. Hall writes, seemed easy in principle. The formidable difficulties involved in converting principles into practice are described by David Waters in his comprehensive account of the solution of the problem of longitude determination. Science provided the basis for the method of "lunars"—in such ways as by the detection of the complex motions of the moon, by the discovery of the speed and aberration of light, and by the computation of the earth's orbital velocity—and it provided the basis for the method of transporting chronometers by the discovery of the isochronism of the pendulum. Instrumental accuracy, stressed by Van Helden and by Waters, was certainly also a telling factor, however; the increased accuracy of telescopes, of pendulum clocks, and of navigational instruments was largely responsible for the perfection of both methods of determining longitude with precision.

Two aspects of Waters's account stand out prominently: the number and complexity of obstacles, both scientific and technical, which had to be surmounted; and the extended length of time—two and one-half centuries—during which scientists and technicians were occupied with the problem. The need became more acute as the exploration of oceans and continents proceeded, and the desire to overcome the difficulties fired the imagination of many scientists. But the solution seemed to come, in Westfall's words, only "after the reservoir of new quantitative science had filled to a certain level."

In the final essay, Richard Olson directly addresses Margaret Jacob's argument that the institutionalization of Newtonianism is intimately related to its supportive role in the ideology of latitudinarian Anglicanism and of Whig political attitudes. If this were the case, he writes, then one should find authors of Tory persuasion and with High Church affiliations expressly attacking Newtonian science. Olson examines the work of three writers who meet his religious and political qualifications: John Arbuthnot, Jonathan Swift, and Samuel Johnson. He finds that they indeed expressed antiscientific sentiments but that they cannot be considered anti-Newtonian. Instead, they were opposed to scientism—the attempt to reduce doctrinal issues and religious or moral problems to the level of scientific questions—and to what they viewed as the excessive pride and apparent insensitivity of scientists. Swift, the most outspoken opponent of science, disparaged *all* science as useless, as did Butler in the seventeenth century; and he was therefore not specifically anti-Newtonian. As an accomplished mathematician, Arbuthnot was in a position to detect unwarranted extensions of valid principles. Johnson was well-read in science and was an ardent supporter of Newtonian science. Olson believes, in conclusion, that High Church Anglicans were generally supportive of Newtonian science while stressing the limitations of scientific knowledge. Those High Church Anglicans who actively pursued science consistently followed Newton's methodology. The ideological role that Newton's world view played in latitudinarian thought is neither "necessary nor sufficient to account for the acceptance and rise to dominance of Newtonian natural philosophy."

We know that philosophical and religious concerns influenced the theoretical aspects of seventeenth-century science and that practical problems motivated the efforts of many scientists. It is another matter, however, to argue that in the formulation of theory religious or political concerns overrode convictions based on observations, experiments, and logic, or that perceived economic or military needs were instrumental in the selection of scientific problems for investigation. To argue this successfully, one is compelled to guess about motivations and to assess their relative importance in the mind of the scientist, a perilous if not impossible task. And, in such a process, the scientist's desire to acquire knowledge for its own sake may be overlooked.

The examples of flaws in contextualist generalizations which have been described in some of the following essays do not, as the authors admit, constitute refutation; but they do serve to question the validity of such generalizations. That "ideological and social factors proved crucial in the development of science in seventeenth-century England"[13] conceals other facets that contributed significantly to scientific advance during this period. Among those treated in the essays are the revolution in scientific instrumentation, the scientific heritage both theoretical and mathematical, and the progress of methodology through criticism and the free exchange of ideas. The success of science in the age of Newton was due to many factors. To assert that ideological and social factors were dominant requires much more evidence and more comprehensive accounts than have yet been presented.

NOTES

1. Steven Shapin, "The Social Uses of Science," in *The Ferment of Knowledge*, ed. G. S. Rousseau and Roy Porter (New York: Cambridge University Press, 1980), pp. 93–139.

2. Ibid., p. 93.

3. Ibid., p. 99. Margaret C. Jacob, *The Newtonians and the English Revolution, 1689–1720* (Ithaca, N.Y.: Cornell University Press, 1976).

4. Ibid., pp. 99–100. J. R. Jacob, *Robert Boyle and the English Revolution* (New York: Burt Franklin, 1977).

5. Ibid., p. 105. J. R. Jacob and M. C. Jacob, "The Anglican Origins of Modern Science: The Metaphysical Foundations of the Whig Constitution," *Isis* 71 (1980): 251–267.

6. Ibid., p. 130.

7. Ibid., p. 132.

8. Ibid., p. 106. B. Hessen, "The Social and Economic Roots of Newton's *Principia*," in *Science at the Crossroads* (London: Kniga, 1931); Robert K. Merton, "Science, Technology and Society in Seventeenth Century England," *Osiris* 4 (1938): 360–632.

9. Sir Isaac Newton, *Principia*, rev. F. Cajori, 2 vols. (Berkeley and Los Angeles: University of California Press, 1962), II:529–530, Proposition XLI, Problem XXII.

10. Jacques Rohault, *Traité de Physique*, 2 vols. (Amsterdam, 1672), I:4–6.

11. For example, some authors lump any seventeenth-century activity involving some degree of rational thought under the word *science*, including under this rubric any empirical attempt to improve a technological product or process. *Newtonianism* is another word that has largely become empty of meaning. It has been variously equated with Newton's natural philosophy, his science, his cosmology, his methodology, his vision of a world of active forces, and his theory of matter (the last divisible into mechanism and materialism).

12. Shapin, "Social Uses of Science," p. 107.

13. Jacob and Jacob, "Anglican Origins of Modern Science," p. 251.

I

THE POETS AND SCIENCE
IN SEVENTEENTH-CENTURY ENGLAND

Earl Miner

Science is so central to modern thought that it is customary to consider what poets have thought about it—whether the poet be John Donne bothered by the new philosophy's calling all in doubt, or Tennyson disturbed over nature red in tooth and claw. To alienate the subject, a procedure that is always revealing, we might ask what a T'ang poet would have thought of scientific conceptions of the nature of poetry. The research of Joseph Needham and others has shown that the Chinese were far ahead of the Europeans in many areas of theoretical and applied science. Yet the Chinese poet would have been amused by the notion that someone we regard as a scientist had anything important to say about so central an enterprise as poetry. If nothing else, this alienation of the topic suggests that in early modern Europe science had not the status it has today. It was natural philosophy. *Science* meant instead, as we all understand, knowledge of all important kinds, and *sapientia religionis* was superior to *scientia* however bravely considered.

What then was the poets' view of science in England during the seventeenth century? During those one hundred years science began to emerge with something like its modern claim on our special interest, and the poets began to realize as much. In what follows I should like briefly to compare some apparent

1

features of world view then and now, and afterward to focus on the lag in the intellectuals' comprehension of and the poets' use of science as an exemplar of their other concerns. I shall close as I began, inquiring why it is of interest to us to know what poets say about science but not what scientists say about poetry.

There are some superficial likenesses between the variously complex intellectual systems of three centuries or so ago and now. Today, the second law of thermodynamics implies entropy. The course of the universe is one of deterioration, or at least of playing out. The course of scientific knowledge, however, is one of progress, since even that law represents an advance on Newtonian laws of inertia. At about 1650 most people also thought of the world as entropic. Human history was a record of deterioration. And yet the true science, the real wisdom, might progress, as even the Reformers looking back to the true church, or *prisca ecclesia*, discovered somewhat by accident.

Of course these resemblances obscure profound differences. The traditional Western view of entropy was associated with the Fall in Eden, after which the race was thought to become shorter-lived, smaller in stature, and worse in important moral characteristics. Wisdom was religious rather than scientific, and the realization of fully adequate truth was generally conceived to belong to the Day of Judgment. Between such views and the modern ones that superficially resemble them, there intervened a variety of ideas about science which slowly came into acceptance. Resistance to them came not only from the poets and divines but also from the politicians, moralists, philosophers, and the as yet unnamed economists and sociologists—as well as, in some ways, the scientists themselves.

The scientific triumphs that culminated in Newtonian physics were mainly, but not solely, in what many term *mechanics*, which I would rather call *dynamics*—a science of motion and of presumed fixed entities, or of "bodies" having, as we would now say, "mass" (with all the questions that either of these words begs). It must also be said that England lagged behind the more advanced continental countries in science and technology in 1650. In fact, as Samuel Butler shows so well in his furious masterpiece *Hudibras*, the England of the middle decades of the seventeenth century was haunted by varieties of learning that were obscurantist: hermeticism, Rosicrucianism, astrology, and others of that ilk. This is not to deny that out of that heady mixture science, modern in some sense and in some kind, struggled into birth. Yet one need only contemplate the titles of the works of Robert Boyle, the beautiful and eloquent nonsense of Thomas Burnet, or Sir Isaac Newton's practicing astrology and alchemy, to understand that the new philosophy shared much with, or was so contaminated (or enlivened) by, the old, among even the most eminent scientists, as to be hardly recognizable by present conceptions of science.

Such familiar but bemusing oddities have numerous implications, among them the slow acceptance by intellectuals then of what today are considered unquestionable scientific advances. It is customary, at least among literary critics, to look on conceptions of the universe as the test case and to speak of a contrast between the old Ptolemaic and the new Copernican universe, of the conception of a geocentric universe yielding to the understanding of a heliocentric one. No such tidy opposition actually existed, however. For one thing, Kepler and Copernicus dressed up their new ideas in some distinctly old garb, perhaps partly for safety's sake but no less because they were unable to shake the new thoroughly free from the old. To my mind, Robert Burton best expresses an intellectual's dilemma in facing claims and counterclaims. The passage is lengthy, I fear, but Burton *is* leisurely, not least in this "Digression of air":

To avoid these paradoxes of the Earth's motion (which the Church of *Rome* hath lately condemned as heretical, as appears by *Blancanus'* and *Fromundus'* writings) our latter Mathematicians have rolled all the stones that may be stirred: and, to solve all appearances and objections, have invented new hypotheses, and fabricated new systems of the World, out of their own *Daedalean* heads. *Fracastorius* will have the earth stand still, as before; and, to avoid that supposition of *Eccentricks & Epicycles*, he hath coined 72 Homo-centricks, to solve [save] all appearances. Nicholas Ramerus will have the Earth the Center of the World, but moveable, and the eighth sphere immoveable, the five upper Planets to move above the Sun, the Sun and Moon about the Earth. Of which Orbs, *Tycho Brahe* puts the Earth the Center immoveable, the stars immoveable, the rest with *Ramerus*, the Planets without Orbs to wander in the Air, keep time and distance, true motion according to that virtue which God hath given them. . . . *Roeslin* . . . censures all, and *Ptolemaeus* himself as insufficient: one [he says] offends against natural Philosophy, another against Optick principles, a third against Mathematical, as not answering to Astronomical observations: one puts a great space betwixt *Saturn's* Orb and the eighth sphere, another too narrow. In his own *hypothesis* he makes the Earth as before the universal Center, the Sun to the five upper Planets, to the eighth sphere he ascribes diurnal motion, Eccentricks and Epicycles to the seven Planets, which [notion] had been formerly exploded; and so . . . as a tinker stops one hole and makes two, he corrects them, and doth worse himself, reforms some, and mars all. In the mean time the World is tossed in a blanket amongst them, they hoise the Earth up and down like a ball, make it stand and go at their pleasures. (*Anatomy of Melancholy*, 2: 65–66)[1]

Burton makes the dilemma plain enough, and the only poet I know of in the seventeenth century to be resolutely Copernican is Sir William Davenant. In the Astragon cantos of his *Gondibert* (bk. 2, cantos 5 and 6), he has his sage, Astragon, discover the heliocentricity of the universe some eight or nine centuries before Kepler and Galileo (canto 5, sts. 17–20). Yet that is mostly by the way. The canto ends by depreciating philosophers, moralists, and historians, and with praise of David's art, poetry.

Other poets could not make up their minds. In fact, they were not entirely sure that the question mattered very much. In *On the Morning of Christ's Nativity*, Milton does use the Ptolemaic system and the music of the spheres as representative of world harmony (see sts. 6–13). The astronomy of *Paradise Lost* is much the same, although Raphael rather pointedly asks Adam:

> What if the Sun
> Be Center to the World, and other Starrs
> By his attractive vertue and thir own
> Incited, dance about him various rounds?
> (8, 122–125)

The idea of world harmony is the important thing, and if the music of the Ptolemaic spheres revolving around the earth be lost, we gain what Dryden so pithily and in a Copernican mood called "The dance of Planets round the radiant sun" (*Fables, Of the Pythagorean Philosophy*, 1. 94). Even the giddy Burton could, like the steady Raphael, pose a Copernican question in that same "Digression":

If our world be small in respect, why may we not suppose a plurality of worlds, those infinite stars visible in the Firmament to be so many Suns, with particular fixt Centers; to have likewise their subordinate planets, as the sun hath his dancing still round him? (2:63)

The need for a harmonious sense of the world was being argued in other, newer terms by scientists themselves. But poets like Dryden were stirred by powerful old conceptions. In *A Song for St. Cecilia's Day*, music in numerous versions—both speculative and practical—accounts for the creation, history, and dissolution of the universe. The close of the first stanza begins by echoing the first two lines of the poem, leading to the combination of the *musica humana* with the *musica mundana*.

> From Harmony, from heav'nly Harmony
> This universal Frame began:
> From Harmony to Harmony
> Through all the compass of the Notes it ran,
> The Diapason closing full in Man.
>
> (11–15)

One cannot tell from these lines whether Dryden was, on balance, a subscriber to the Ptolemaic or Copernican view, and I doubt that he knew, either. Faith in a harmonious and providential view of history mattered more than the "philosophical" (i.e., scientific) issue whether the earth or the sun lay at the center of the universe.

This most familiar of scientific issues from the century obscures others. One reason intellectuals lagged in understanding science was that they lacked the mathematical skills we would expect of a high school graduate. We need not ask what the poets knew about calculus, to which Descartes and others were making important contributions; but we may ask about algebra, which the Arabs had made plain centuries earlier. I find no reference to it in important poets before the Restoration, except for Ben Jonson in *The Alchemist*. Shakespeare and even Milton (d. 1674) are silent on the matter. To Butler in *Hudibras*, it is a kind of arcanum as simply unnecessary and probably ridiculous as the other accomplishments of his hero. After reviewing Hudibras's achievements in logic and rhetoric, Butler next regales us with this:

> In *Mathematicks* he was greater
> Then *Tycho Brahe* or *Erra Pater*:
> For he by *Geometrick* scale
> Could take the size of *Pots of Ale*;
> Resolve by Sines and Tangents straight,

If *Bread* or *Butter* wanted weight;
And wisely tell what hour o'th' day
The Clock does strike, by Algebra.
                        (I, 1, 119–26)

In a well-known parenthesis in his preface to *Fables,* Dryden speaks of "Mr. Hobbes . . . (studying poetry as he did mathematics, when it was too late)."[2] And indeed Hobbes did make a fool of himself by arguing that he knew how to square the circle. We hear another tone, however, in *Religio Laici,* where Dryden deals with the complexities of biblical hermeneutics, "which he who well can sort / May afterwards make *Algebra* a Sport" (238–239).[3] A century before, in translating *The Elements of Geometrie* (1570), Sir Henry Billingsley could speak in his preface of "that more secret and subtill part of Arithmetike, commonly called Algebra." A mathematical art such as algebra remained hidden and difficult, especially for men whose boyhoods were passed in educational systems devoted to teaching things other than those discussed at Gresham College.

Another example of the slow acceptance of new scientific knowledge may be found in the vicissitudes of atomism. In atomism we have an issue that was to have enormous implications much later, in modern physics, but which to the mind of early modern Europe was definitely a matter of the past and present. Bacon of course rejected atomism, and to divines contemporary with him atomism was a pagan, Epicurean, heresy. They held to the four elements—heat, cold, moisture, and dryness—as constituents of matter. Therefore, as classical atomism was revived, its proponents necessarily had to deal with those four elements in some way. By 1656 the eminently respectable John Evelyn, F.R.S., had translated the first book of Lucretius. By 1682 Thomas Creech brought out the whole of that author's work in English. This revival coincided with Continental influences, including such thinkers as Gassendi. In 1654 Walter Charleton sought to "solve appearances" with a compromise between elemental and atomic theories of matter. In the title of his work *Physilogia Epicuro-Gassendo Charletoniana: Or a Fabrick of Science Natural, Upon the Hypothesis of Atoms* (London, 1654), we find a very early use of *science* in the modern sense; and in the following passage we discover his accommodation: "Patrons of Atoms do not . . . deny the Existence of those four Elements admitted by most Philosophers: but allow them to be *Elementa Secundaria*, Elements Elementated, *i.e.*, consisting of Atomes, as their First and Highest Principles" (p. 100).

It is attractive to find a harbinger of modernity in this, but it is quite consonant with providential history, as Charleton would have been the first to say. Dryden followed his friend Charleton in presuming a divine creation and the priority of atoms. We see both features in *A Song for St. Cecilia's Day*, when God's tuneful voice summons atoms prior to the four elements.

> When Nature underneath a heap
>   Of Jarring Atomes lay,
>   And cou'd not heave her Head
> The tuneful Voice was heard from high,
>   Arise ye more than dead.
> *Then* cold, and hot, and moist, and dry,
>   In order to their stations leap.
>   (3–9; stress added)

No doubt some scientific progress can be seen in this. At least it is clearer than Milton's passage on chaos in *Paradise Lost* (1667):

> For hot, cold, moist, and dry, four Champions fierce
> Strive here for Maistrie, and to Battle bring
> Thir embryon Atoms; they around the flag
> Of each his Faction; in their several Clanns,
> Light-arm'd or heavy, sharp, smooth, swift or slow,
> Swarm populous.
>   (2, 898–903)

By the end of the century the four elements were more or less assigned to that scientific rubbish heap where phlogiston lies. As Robert Hugh Kargon has put it, "By 1700, atomism as a mechanical philosophy was, in England, the conservative view."[4] Of course protons, neutrons, antimatter, and quarks, charmed or otherwise, lay far in the future.

To such an extent did scientific ideas make slow headway in seventeenth-century England. The slowness is as indisputable as the headway. Having emphasized the slowness, however, I owe it to the century to acknowledge the headway. This is exemplified by the decline in the belief in magic. The so-called Wizard Earl of Northumberland, Henry Percy, who succeeded as the ninth earl in 1585, gathered about him an intellectual circle that included poets and dramatists. They seriously considered the numerological symbolism of John Dee, the Copernicanism of Dee's student Thomas Digges, the new experimental emphasis of Gilbert, and the atomism combined with Neoplatonism and Pythagoreanism of Giordano Bruno.[5] With great effort the next century separated the magical from the genuinely scientific. For, although the two decades from about 1640 to 1660 shared with the 1580s a great revival of natural magic, earlier ideas of the occult, of astrology, and of the magician would gradually disappear.

We may think that Shakespeare's handling of magic involves only plot and dramatic characterization, or that his Welsh magic and Scottish witches signify artistic convenience, much as such elements in Shadwell's *Lancashire Witches* later served to provide dramatic atmosphere. In *The Tempest*, however, Prospero is a genuine wizard of the best European extraction, and his magic is not compromised by the diminished scale of the fairies in *A Midsummer Night's Dream*. Yet even Prospero puts aside his magic to return from an island to his proper society. Thereafter, dramatic treatment of magic became the full possession of only the masque; comparable interests are satiric in Jonson's *Alchemist* and *Magnetic Lady*. A concern with witches did linger: James I was mightily interested, and Sir Thomas Browne argued for the existence of evil spirits as a necessary corollary to good ones (*Religio Medici*, pt. 1, sec. 30).[6] Milton's *Comus* (1637–1645, in its three versions) all but marks the end of magic in English literature, or at least of magicians whose role is essential to the credibility of the plot. Thereafter we find some retrospection in plays, in what Dryden termed "the fairy way of writing"; but by the end of the century the magician is as dead as Pan.

The slow adjustment by English writers to science in the seventeenth century looks a bit different if we consider the matter in another light, that of science as a representative concern or emblem. In this regard the *locus classicus*, or the jaded and misunderstood example, is that of Donne's first *Anniversary*, occasioned by the death of Elizabeth Drury.

And new Philosophy cals all in doubt,
The Element of fire is quite put out;
The Sunne is lost, and th'earth, and no mans wit
Can well direct him, where to looke for it.

(205–208)[7]

Obviously it is the new astronomy which is referred to, since it leaves intellectuals uncertain (as it did Burton) whether the earth or the sun is at the center of the universe. Yet what precisely (not "all") is called in doubt? For one thing, there is the constitution of the universe. That is only the grounds, however, for the peril to other valued assumptions:

'Tis all in pieces, all cohaerence gone;
All iust supply, and all Relation:
Prince, Subiect, Father, Sonne are things forgot,
For every man alone thinkes he hath got
To be a Phoenix.

(213–217)

In short, the problems are social and moral. The system of obedience of subjects to princes and of sons to fathers is breaking down. The great chain of being, so romanticized by literary critics, is threatened by the loss of "degree," the principle by which a few rule and the many obey. We observe that Donne does not mention women or daughters. It is an authoritarian, patriarchal system of order which the new philosophy calls in doubt.

Such doubts probably did trouble Donne in a scientific sense, as they certainly did Burton, but science was for him an exemplar of larger issues. Burton demonstrates this view as well in his argument in the long initial epistle to the reader. At one point the argument reaches a climax over Copernicus, who "is of opinion the earth is a planet, moves, and shines to others," and so forth (1:86). This is the argument from wholes, but he has also dealt with particulars, and he sees them everywhere, as in this paragraph of expostulations ending a number of "To see" paragraphs:

To see wise men degraded, fools preferred, one govern towns and cities, and yet a silly woman over-rules him at home . . . To see horses ride in a coach, men draw it; dogs devour their masters; towers build masons; children rule;

old men go to school; women wear the breeches, sheep demolish towns, devour men, &c, and in a word, the world turned upside downward! *O viveret Democritus!* (1:73)

Once again we observe that astronomy is only an exemplar of the larger context of social disorder. The debates over the universe merely provide one model of a feared loss of the parental, male-dominated, hierarchical scheme of social order.

Later English writers were no less concerned with science as exemplar, but their attention found different focus. We have seen that Milton was concerned with a harmonious world order. In fact, the passage quoted earlier to show Milton's tolerance of the Copernican view comes in a somewhat nervous parenthesis in which Raphael lectures Adam. The sociable archangel almost begins with the words, "whether Heav'n move or Earth, / Imports not" (8, 238); and he concludes with the advice: "Be lowlie wise" (8, 173), that is, prefer religious wisdom to philosophical knowledge. The organization of the universe merely exemplifies divine providence and magnificence. The question of geocentricity, like Burton's earth itself, is ultimately neither here nor there.

Samuel Butler was not so much nervous as furious over matters that did not matter. He had one predecessor, that short-fused Scot, Alexander Ross. Ross has been maligned by people who suggest that he opposed ideas simply because they were true. Actually, he objected to anything that was new. In 1634 his *Commentum de Terrae Motu Circulari* opposed the heliocentric view. After Wilkins's great *Discourse concerning a New Planet*, he attacked with *The New Planet no Planet* (1646). Sir Kenelm Digby's crazy treatise on sympathetic cures brought forth a Rossian diatribe in *The Philosophicall Touch-Stone* (1645). Also in 1645, his *Medicus Medicatus* rose up against Digby and Browne's *Religio Medici*. His *Arcana Microcosmi* (1651, 1652) has the distinction of attacking Bacon, Browne's *Vulgar Errors*, and Harvey's *De Generatione*. He excoriated Hobbes in his *Leviathan drawn out with a Hook* (1653). Later he branded the impostures of the *Koran*. Anything new—anything—was wrong, and like the rain, Ross falls on the just and the unjust. Since these works saw publication one after the other, and since his *Mystagogus Poeticus* (1647) was also highly popular (it still helps us understand Milton and Dryden), we can only assume that he was a spokesman for many people disturbed by new thought, definitely a man for the age if not for all time.

Ross seems irrational, even crazy, to us. Samuel Butler brought a *furor rationis* that held science an exemplar of useless learning, a view that was to last on, as the third voyage of *Gulliver's Travels* well shows. Butler's terrible, terrible, great poem *Hudibras* gives no intellectual quarter. In his second canto he begins by dismissing Empedocles and Ross in a characteristic, anachronistic couplet: "There was an ancient sage *Philosopher*, That had read *Alexander Ross* over" (1, 2, 1–2). The differences of centuries vanish. Nonutilitarian nonsense remains despicable, and in that sense an ancient Greek philosopher can be said to have read the truculent Ross. Anything smacking of the speculative, the purely intellectual, was branded "mystick" by this poet, the ancient with the modern, the good with the bad. If he had any special aversion, it was to what he terms the "mystick Learning" of those magical middle decades of the century, when rebellion, nonsense, degradation, and useless learning went together. Here is part of his furious characterization of Hudibras's second, Ralph, the degraded Sancho Panza to that degraded Don Quixote. Ralph is one

> For mystick Learning, wondrous able
> In Magick, *Talisman*, and *Cabal*,
> Whose primitive tradition reaches
> As far as *Adam*'s first green breeches:
> Deepsighted in Intelligences,
> Idea's, Atomes, Influences;
> And much of *Terra Incognita*,
> Th'Intelligible world could say:
> A deep occult Philosopher,
> As learn'd as the *Wild Irish* are,
> Or Sir *Agrippa*, for profound
> And solid Lying much renown'd:
> He *Anthroposophus*, and *Floud*,
> And *Jacob Behmen* understood;
> Knew many an Amulet and Charm,
> That would do neither good nor harm:

In *Rosy-Crucian* Lore as learned
As he that *Verè adeptus* earned.
                                    (1, 1, 523—540)

It is not surprising that Ralph also claims to foretell public events by astrological means (see 567—574).

Nor should we be astonished that Butler includes a proper astrologer, Sidrophel, among his despicable characters. The argument to canto 3 in the second part depicts that star-lover Sidrophel as "*the* Rosy-crucian." In the general mess of the poem, he therefore shares much with Ralph. But he is not a political fanatic. He is merely a charlatan, that is, a kind of scientist.

Some remarkable things happen with Sidrophel. The most important for our purposes has not become a hackneyed example as has Donne's passage on the new philosophy. At the end of the second part Hudibras the anti-hero addresses a spoilt "heroical epistle" to Sidrophel. Ovid had established the heroic epistle as a woman's dignified address to her lover in a strained or tragic situation. In Butler's sarcastic letter, the chronology of the poem moves from the 1650s to the Restoration.[8] This shift in time does not bring Reason to her throne, however. Instead, Sidrophel becomes a virtuoso of the Royal Society. In this, one of the most extraordinary and telling metamorphoses in English poetry, Butler depicts the new science as a timeless example of irrationality and uselessness, given contemporary guise in equally nonsensical, impractical terms. It is sheer theory without utility.

Butler's transformation of Sidrophel is extraordinary. The charge against science is not confined to him, however: there are also the Laputan madness in *Gulliver's Travels* and the hack's modernity in Swift's *Tale of a Tub*. The fear of scientific modernity, the distrust of knowledge for its own sake, is also shown in Thomas Shadwell's best play, *The Virtuoso*. The title implies a member of the Royal Society or some other of the gentleman or professional scientists. Sir Nicholas Gimcrack is discovered on a table, a string in his mouth connected to a frog in water. The frog's motions teach Sir Nicholas the theory of swimming. He never seeks to swim in water, which would be practical. As he says, "I content myself with the speculative part of swimming, I care not for the Practick. I seldom bring any thing to use, 'tis not my way, Knowledge is my ultimate end" (act 2).[9] The accusation is clear: scientific knowledge is useless, a thing that at best, in Butler's phrase, "would do neither good nor harm."

In all this we see an older counterpart of one current image of the scientist, that of a spooky figure who may do this or that mysterious but who is basically ridiculous, of no use. At heart this is a species of antiintellectualism. To any true scientist, or true admirer of any kind of knowledge, knowledge is first of all valuable for intrinsic reasons, for its own sake. Butler, Shadwell, and Swift are each of them important writers, but surely they have got something seriously wrong. For if their charge against science holds, the same charge can be turned upon them: their own art of poetry is no more useful than science. The one valuable thing in this is that it brings into focus an important issue, one that the solely utilitarian imperative distorts—the nature of knowledge and its value for its own sake. However sympathetic to the utilitarian imperative any of us may be, we must reject its reductive application. It is a relief therefore to turn to Dryden's quite different use of science as an exemplar.

Among Dryden's views of history—for like any person he had more than one and they changed over the years—is a belief in the possibility of human progress. He saw this progress as brought about by individuals' great achievements in human knowledge, which enable a given age to advance on its predecessors. In the epilogue to the second part of *The Conquest of Granada*, Dryden claimed that he and his contemporary dramatists had made some improvements on the drama of the English past. He was attacked for this; and so in due course he wrote a "Defence" of that epilogue, in which he remarks: "I profess to have no other ambition in this essay than that poetry may not go backward, when all other arts and sciences are advancing" (Watson, 1:169). With this profession we have arrived in a new intellectual sphere. Those "arts and sciences" are now exemplary of what is good; and poetry—or drama—may take heart from them. Attack and sarcasm have vanished; leaving hope in their stead.

Dryden's declaration in his own voice gives special urgency

to a particularly significant moment in his earlier *Essay of Dramatic Poesy*, a moment I like to refer to as Crites' Question. Here is the rhetorical query by Crites, one of the four speakers in the dialogue.

Is it not evident in these last hundred years (when the study of philosophy [i.e., science] has been the business of all the virtuosi in Christendom) that almost a new nature has been revealed to us? that more errors of the school [i.e., of scholasticism] have been detected, more useful experiments in philosophy have been made, more noble secrets in optics, medicine, anatomy, astronomy discovered, than in all those credulous and doting ages from Aristotle to us? so true is it, that nothing spreads more fast than science, when rightly and generally cultivated? (Watson, 1:26)

Dryden allows for use as it deserves, but his *intellectual* excitement is remarkable by contrast with his predecessors, or even with such as Shadwell and Swift later on. The passage is also one to give heart to historians, whether of literature or of science. For in his talk of "ages" here and elsewhere, Dryden gives birth in English to one of the basic concepts of subsequent historical thought, the concept of an age or period.[10] Dryden offers hope for intellectual inquiry in itself, a historical scheme to make it intelligible, and a vision of science as an exemplar for what may be gained in other kinds of knowledge.

Dryden gave Crites' Question a series of affirmative answers in his poetry and prose. His most confident statement for the collective human enterprise came during the reign of Charles II. In *Annus Mirabilis* he introduced his optimism for science in what was to be his basic mode of affirming faith in the future, the progress piece or *translatio studii*, an account of the progress of knowledge of a given kind. Because his important poetic predecessors had consistently adhered to the theory of the decay of nature, historical entropy, his progress pieces are the more remarkable. Here are the last two stanzas of the progress piece on navigation in *Annus Mirabilis*, with their prediction that England would discover the means of determining longitude and of becoming a maritime imperial power, displacing the Dutch.

> The Ebbs of Tydes, and their mysterious flow,
>   We, as Arts Elements shall understand:
> And as by Line upon the Ocean go,
>   Whose paths shall be familiar as the Land.

Instructed ships shall sail to quick Commerce;
        By which remotest Regions are alli'd:
Which makes one City of the Universe,
        Where some may gain, and all may be suppli'd.
                                        (645–652)

If the scientific premise were not already clear in this, he goes
on to apostrophize the Royal Society.

This I fore-tel, from your auspicious care,
        Who great in search of God and Nature grow:
Who best your wise Creator's praise declare,
        Since best to praise his works is best to know.

O truly Royal! who behold the Law,
        And rule of beings in your Maker's mind,
And thence, like Limbecks, rich Idea's draw
        To fit the levell'd use of humane kind.
                                        (657–664)

With the personal shocks of the 1688 Revolution, and even
before that in his conversion to Catholicism, Dryden found it
necessary to retreat from his optimism about science and in
fact to question his faith in the future. But he never entirely
lost either his optimism or his faith, because he was able to base
them on his religion and on his art. In his retreat to what
mattered most to him, he unquestionably grew less optimistic
about secular and scientific promises. As he weathered various
shocks, he came to resemble his earlier contemporaries more,
although he never wholly abandoned his hopes and continued
to invest them in individuals—in, for example, the *lexis* and
praxis of that Homer whom Dryden thought he might translate even as death overtook him.

On balance, the Dryden of Crites' Question shows the far-
thest advance that poetry made in understanding science during the seventeenth century in England. Above all, he had a
historical vision that sets him apart from other poets and critics
in the English tradition. To use the kind of metaphor he
would have used, we might say that if others were to see
farther than he, they still looked through his telescope. It was
not he but later writers who (like earlier ones) mounted attacks
on science, against such supposed enemies as those Blake
attacked in terms of the atoms of Democritus and Newton's

particles of light. The discovery of the two cultures was not Dryden's, and for that matter it was not that of his century. If writers like Donne, Milton, and Dryden had a fault, it was their inability to distinguish adequately, as we see things, the separate provinces of science and literature. Yet the scientists of the seventeenth century were no better at making our distinctions. The history of science after that century is one of attempts to shake off poetics and metaphor, to secularize, and above all to specialize ever more narrowly. The direction of what is often termed classical or Newtonian science has clearly been toward disentanglement from what Butler excoriated as "mystick Learning."

It cannot be maintained that poetry today has so radically departed from the "mystick" or that it has grown so specialized and arcane. This brings me back to my initial question, why we should find it of interest what poets think of science but not of what scientists think of poetry. We have seen that languages of science, such as algebra, were thought difficult by seventeenth-century poets; and modern poets, along with literary critics such as myself, still cannot claim to understand the languages of science at all adequately. By contrast, poetry retains versions of natural language that are much more accessible to any scientist who wishes to understand. Is this not a paradox—much like that of my work appearing among the work of historians of science?

The explanation of this seeming paradox appears to me to involve the differing cognitive statuses of literature and science. Within the boundaries of ordinary understanding and for the purposes of the versions of ordinary language typical of literature, science is simply necessary to give a grounding in reality. Keats provides an example of this in his sonnet "On First Looking into Chapman's Homer," an experience described in the conclusion of the poem by two similes:

> Then felt I like some watcher of the skies
>   When a new planet swims into his ken;
> Or like stout Cortez when with eagle eyes
>   He star'd at the Pacific—and all his men
> Look'd at each other with a wild surmise—
>   Silent, upon a peak in Darien.[11]

He has got the discoverer of the Pacific and the place of discovery wrong. No geographer, no historian, who made such errors would still be appreciated today as Keats is. To such an extent do poets require science, even if they err in major particulars. Poets may do so, in my view, because their art shares with the other arts what I would term a virtual—neither true nor false—cognitive status, as opposed to the predictive status of scientific knowledge.[12] We now know that creativity in science greatly resembles that in the arts. Since Gödel and Heisenberg even literary critics are aware of the problematics of scientific knowledge. But modern science has disencumbered itself from the poetic, religious, and "mystick" attributes it possessed during the seventeenth century, and it constantly aspires to predicate what truly is.

To put it another way, science does not now depend in any discernible fashion on these "mystick" kinds of knowledge. Science is simply more independent, purer. By the same token, however, scientists are less able to include the knowledge and values of the arts than literature can those of science. If a scientist appreciates poetry—or more likely, music, with its mathematical affinities—it is in a private, not a scientific, capacity. As the sciences have become purer, they have grown ever more remote from ordinary, from literary, understanding. Scientific knowledge may be adaptable from one version to another, as the remarkable recent advances in biochemistry, for example, have shown; but the purest science, mathematics, has become less and less intelligible, and this well illustrates a cultural dilemma.

The dilemma is not one of the two—or more—cultures, although mutual ignorance between poets and scientists, and the students of each, may indeed be an intellectual scandal. Much worse is that scientists simply cannot use poetry, and that poets still need science to write though they understand less and less of what scientific knowledge has become. Perhaps there is some full and happy solution to this problem which escapes me. It seems more likely that, like other important human problems, this one requires recognition when solution lies beyond our means. My aim in this is to be skeptical, not pessimistic or cynical. Surely we can at least understand what the problem is. We can inquire into the differing cognitive statuses of such now familiar specialties as science, poetry, and history. And since each of these has its history, it seems to me that we must rest our hopes on historical study of what are, after all, only different kinds of human knowledge.

NOTES

1. Burton is cited in the text by volume and page in *The Anatomy of Melancholy*, ed. A. R. Shiletto, 3 vols. (London: George Bell, 1896).

2. Preface to *Fables Ancient and Modern* in John Dryden, *Of Dramatic Poesy and Other Critical Essays*, ed. George Watson, 2 vols. (London: Dent, 1962), 2:275; and hereafter in the text as Watson. The edition used for the poems is *The Poems of John Dryden*, ed. James Kinsley, 4 vols. (Oxford: Clarendon, 1958). The text used for Milton is *The Works of John Milton*, ed. Frank Allen Patterson et al., 18 vols. (New York: Columbia University Press, 1931–38), 2, Part 1, p. 240; that for Davenant, *Sir William Davenant's Gondibert*, ed. David F. Gladish (Oxford: Clarendon, 1971), pp. 153–154; and that for Butler, Samuel Butler, *Hudibras*, ed. John Wilders (Oxford: Clarendon, 1967), pp. 4–5.

3. Dryden's poetry is quoted from *The Poems of John Dryden*, ed. James Kinsley, 4 vols. (Oxford: Clarendon, 1958), 1:317.

4. Robert Hugh Kargon, *Atomism in England from Hariot to Newton* (Oxford: Clarendon, 1966), p. 133. This study has an excellent history of science bibliography, as do certain publications sponsored by the Clark Library. These include: A. Rupert Hall, "English Scientific Literature in the Seventeenth Century," in *Scientific Literature in Sixteenth & Seventeenth Century England* (Los Angeles: Clark Library, 1962); Ladislo Reti, "Von Helmont, Boyle and the Alkahest," in *Some Aspects of Seventeenth-Century Medicine and Science* (Los Angeles: Clark Library, 1969); Robert S. Westman, "Magical Reform and Astronomical Reform: The Yates Thesis Reconsidered," in *Hermeticism and the Scientific Revolution* (Los Angeles: Clark Library, 1977); and C. D. O'Malley, "The English Physician in the Earlier Eighteenth Century," in *England in the Restoration and Early Eighteenth Century*, ed. H. T. Swedenberg, Jr. (Berkeley, Los Angeles, London: University of California Press, 1972). Work by Thomas Kuhn and entries in the *Dictionary of Scientific Biography* yield further valuable information of theoretical and historical kinds. My proper subject involves more directly the response of writers to science, and for that topic there are useful bibliographies in: Douglas Bush, *English Literature in the Earlier Seventeenth Century*, 2d ed. (Oxford: Clarendon, 1962), pp. 507–511; and James Sutherland, *English Literature of the Late Seventeenth Century* (Oxford: Clarendon, 1969), pp. 489–492. Numerous other studies exist, of which those by Marjorie Hope Nicholson and Joseph A. Mazzeo are perhaps closest to what is dealt with here.

5. Kargon, *Atomism in England*, pp. 6–12. What follows concerns certain matters treated somewhat differently by Keith Thomas in his *Religion and the Decline of Magic* (New York: Scribner, 1971).

6. *The Works of Sir Thomas Browne*, ed. Geoffrey Keynes, 4 vols. (Chicago: University of Chicago Press, 1964), 1:40–41.

7. The text used for Donne is *John Donne: The Anniversaries*, ed. Frank Manley (Baltimore: Johns Hopkins University Press, 1963). The quotations given are from pp. 73–74.

8. On the chronology of *Hudibras*, see my *Restoration Mode from Milton to Dryden* (Princeton: Princeton University Press, 1974), pp. 163–174.

9. The text used for Shadwell is *The Complete Works of Thomas Shadwell*, ed. Montague Summers, 5 vols. (reprint ed., New York: Benjamin Blom, 1968), 3:127.

10. I have discussed this elsewhere, most recently in "The Poetics of the Critical Act," in *Evidence in Literary Scholarship*, ed. René Wellek and Alvaro Ribeiro (Oxford: Clarendon, 1979), pp. 47–48.

11. Quoted from *An Oxford Anthology of English Poetry*, ed. Howard Foster Lowry and Willard Thorp (New York: Oxford University Press, 1935), p. 827.

12. For a discussion, see Earl Miner, "That Literature is a Kind of Knowledge," *Critical Inquiry* 2 (1976): 487–517.

II

OLDENBURG, THE *PHILOSOPHICAL TRANSACTIONS*, AND TECHNOLOGY

Marie Boas Hall

Conventionally, and because it is valid for more recent times, historians have interpreted the interaction of science and technology in terms of the application of science to technology. But not only is it the case that Carnot could not have written his classic work without the steam engine; there is good reason to believe that in the seventeenth century many men looked to a study of the arts and crafts primarily to gain insight into the workings of nature. As Walter Houghton pointed out forty years ago in his classic study "The History of Trades: Its Relation to Seventeenth Century Thought," the Baconian vision of the history of trades had two faces: it could mean, simply, the study of technology to improve utility, or it could more fruitfully mean the study of technology to assist the development of science.[1] My aim in what follows is to examine the early attitude of the Royal Society to the uses of technology, especially as this is exhibited in the *Philosophical Transactions*, founded in 1665.

Now it must be noted that the *Philosophical Transactions* do not entirely supply the key to the Royal Society's attitude, for they were not, in the seventeenth century, the Royal Society's journal, but rather a private venture, originally conceived by the Royal Society's industrious secretary, Henry Oldenburg. The scope of the journal is indicated by the full title, *Philosophical Transactions: Giving some Accompt of the Present Undertakings, Studies, and Labours of the Ingenious in many Considerable Parts of the World*, which plainly shows that Oldenburg did not intend to confine the contents to the activities of the Royal Society alone, nor even to what was being done in England. This being so, it is necessary to consider what Oldenburg's own attitude to technological matters was when he began to edit his new journal. Yet inasmuch as he had been a Fellow of the Royal Society for over four years when he began his editorship, and as moreover readers both in England and abroad quickly identified the *Philosophical Transactions* with the Royal Society—of whose activities they were the only public source of news and with whose philosophy they were so closely identified—it is necessary also to consider the Royal Society's own attitude to technology. (In fact, Oldenburg fought a losing battle to establish the journal as his own, not the Society's, in the eyes of his readers.)[2]

As Oldenburg himself is so much less well known than his journal, it may be helpful to begin by sketching Oldenburg's intellectual background and the origins of his own interest in technology, before considering the attitude of the Royal Society of which he was both a member and an officer. This is relevant not only to his later expression of these interests in his journal but also as exemplifying the process whereby men of his generation could acquire such interests. His origins were conventionally professional: he was born in Bremen in about 1618, the son of a teacher in the Bremen Gymnasium who later moved to a minor university.[3] His whole family was of the academic and legal class, and he was educated (in the gymnasium) to follow them, graduating in 1639 with a master of arts in theology after a thorough grounding in the liberal arts. Like most ambitious German youth of his day he decided to tour Europe; he began in the Low Countries (he had introductions to professors there); later he traveled more widely and became a notable linguist. He seems to have earned his living by acting as tutor to various wellborn boys and young men, some certainly English—he may even have visited England, but these years are obscure. In 1653 he emerges again into history as an envoy of the city of Bremen to Cromwell, sent by the Senate of Bremen to negotiate for protection of the city's shipping against depredations during the Anglo-Dutch wars. He proved himself a successful diplomat (he was to use these talents to advantage later in life) and was complimented by no less a person than John Milton on the excellence of his spoken English. He rapidly became friendly with Milton and with many of Milton's circle. Of those who influenced his later life the most important were, the pious, learned, and beautiful Lady Ranelagh, through whom he met Robert Boyle, her youngest brother; Samuel Hartlib, German by origin, English by choice, an indefatigable supporter of projects for the good of mankind; and John Dury, an equally indefatigable propagandist and salesman for Christian unity, whose then baby daughter Oldenburg was later to marry. These, and many more, clearly found him sympathetic and a worthy member of the Parliamentarian community.

Oldenburg seems to have found England congenial, but he needed paid employment. After visiting a number of his former pupils, he went to Oxford in the spring of 1656, partly as a tourist, partly seeking employment as a tutor. Perhaps through Boyle he secured an introduction to John Wilkins, then warden of Wadham, and met others of the club of young men of mixed political and religious leanings who espoused the new learning. As he wrote to a theological friend abroad,

I have begun to enter into companionship with some few men who bend their minds to the more solid studies rather than to others, and are disgusted with Scholastic Theology and Nominalist Philosophy. They are followers of Nature itself, and of truth.[4]

Oldenburg's excitement was genuine; impressed by the potentialities of the new approach he endeavoured to acquire the necessary understanding of the interests and methods of his new acquaintances, rapidly becoming not merely an admirer of the new philosophy but an able student and exponent of it. Though never a natural philosopher himself, he quickly demonstrated that he could learn first to follow and then to judge a scientific discussion, argument, and investigation, and he must at this time if not earlier have made himself familiar with Bacon's ideas. All unconsciously he was preparing himself for his future role as secretary of the Royal Society.

He still had his living to earn, however, and here he was fortunate in the patronage of Lady Ranelagh and of Boyle. He was soon appointed tutor to Richard Jones, Lady Ranelagh's son, and in the summer of 1656 set off with his pupil on the grand tour, not to return until the eve of Charles II's triumphal entry into London in May 1660. No doubt young Jones benefited from his instruction and travels, but he cannot have benefited nearly as much as did his tutor. Because Boyle had already become his patron, Oldenburg took care to search for news to send to Oxford which would interest that noble scientist. As a means to this end he sought contact with intellectual circles he would earlier never have tried to enter and learned to appreciate the French version of the New Learning with its strongly Cartesian approach—especially after his sense of duty relaxed enough to permit his rather idle pupil to go to Paris, where the latter learned the arts of polite society and his tutor attended the scientific academies.

All this taught Oldenburg how to assimilate and pass on scientific and learned news, how to manage an intellectual correspondence, and how to keep a record of such a correspondence for future use.[5] (The last point explains why we know now what subjects he explored.) During this period Oldenburg's two chief correspondents were Robert Boyle and Samuel Hartlib. Both had clearly requested news from abroad, and both wrote with questions about what had been sent, though Hartlib was so full of news of his own circle as to be less concerned with what Oldenburg was learning. For Boyle, Oldenburg reported on the general intellectual scene, the existence of clubs, societies, salons, and academies, first in the provinces (notably from Saumur in the Loire Valley and Castres in the Midi) and later in Paris. His entree into these circles came largely from the social status of his pupil (though the young English milord clearly enjoyed the dancing and riding academies rather than those devoted to natural philosophy which so interested his tutor); but he also endeavoured to secure introductions from one town to another, so that he might meet chemists, alchemists, natural philosophers, and physicians.[6] As time went on, Oldenburg, his native shrewdness stimulated both by contact with new ideas and by questions from Boyle, sent fewer accounts of wonderful remedies, *magnes aquae*, and alchemical claims, and wrote more about theory and experiment. This is most apparent after the period of his arrival in Paris in 1659 and his entry into the Montmor Academy, whose Cartesianism instructed him in the potentialities of theoretical natural philosophy, though the Baconianism he had imbibed in Oxford kept him from becoming a doctrinaire Cartesian. What he had to report was helpful to Boyle, and his standing with the Montmor academicians helps explain his later entry into the Royal Society.

What Oldenburg recounted to Hartlib was necessarily very different because, whatever had been the case in 1646, by 1656 Boyle's and Hartlib's interests were very different. Both still called themselves Baconians, but what they meant by this was poles apart, as Oldenburg perceived. Hartlib's view of Baconianism was a limited and practical one: the "relief of man's estate" meant to him useful invention, craft, agricultural technology, simple aids to everyday affairs, and publicity of all these so that men might benefit immediately.[7] It is, I think, significant that at the time when Oldenburg was writing of the intellectual excitements of Oxford Baconianism to some of his correspondents, he wrote to Hartlib about Wilkins's glass beehives.[8] Later, traveling in France and Germany, he described to Hartlib practical matters only: medical remedies, chemical, optical and mechanical inventions, practical and religious publications. He knew very well that to Hartlib chemistry or alchemy meant the preparation of new remedies and optics meant better lighting.[9] Improved agriculture, new clocks, new lens-grinding machines were ends in themselves rather than instruments for aiding the investigation of nature. Hartlib was not interested in trying to understand how these things worked, nor in what principles were involved; to him they were admirable novelties *tout court*. So lenses for spectacles might rate higher on the scale of utility than lenses for telescopes, and the introduction of fruit trees to new areas higher than any agricultural theory. Did Hartlib understand that Cressy Dymock's new plough was not in the long run as important in the scheme of things as Boyle's air pump, even though the first was but a project and the second was an instrument already in existence which opened intriguing new vistas into nature? One may doubt it, especially in the light of Oldenburg's remark to him from Paris—"Seeing you care not for ye philosophical discourses of our Clubs, you are not to looke for many real Experiments from Frenchmen."[10] To be sure, Hartlib would have liked more news about French inventions of chariots and other useful things, and he was happy enough with the news of chemical Germans and practical Dutch artisans which Oldenburg sent him. Oldenburg clearly showed considerable sympathy with and understanding of French gardening and similar matters, news of which he sent to Hartlib for the delectation of Hartlib himself, of John Evelyn (whom Oldenburg had not yet met), and of John Beale.[11]

It is reasonable to assume that it was a combination of Oldenburg's skill as a correspondent (as appreciated by Boyle and Evelyn) and of his familiarity with the Paris "Clubs" and their promotion of the New Learning that made it natural that, little more than six months after his return to England, he should be proposed a member of the nascent Royal Society, "composed of extremely learned men, remarkably well versed in mathematics and experimental science," as he described it before he knew he was to be one of this assembly.[12] In fact, his name was suggested at the first, organizational meeting of 28 November 1660, and a month later (26 December) he was proposed a candidate, in spite of his foreign nationality.[13] He was an active participant in the Society's meetings, was put on committees (one "for considering of proper questions to be inquired of in the remotest parts of the world" on 6 February 1660/61, and one "for erecting a library, and examining the generation of insects" on 15 May 1661),[14] and was named one of the Original Fellows in the charters of 1662 and 1663. What is more, he was named as one of the two secretaries of the Society in both charters, a position he must have owed in large part to Boyle's patronage, though clearly his skill in correspondence and wide acquaintance helped to recommend him to the leading members of the Royal Society. He was to fill this post with zeal and ability until his death in September 1677—attending meetings, keeping the minutes, the register books, and the letter books, writing the letters ordered, managing a vast correspondence whose salient points he communicated at meetings, and seeing to a large part of the day-to-day running of the Society.

This Society, which began with the aim of "promoting of experimental learning," as the first form of subscription put it,[15] rapidly showed its interest in Baconian "histories of trades" and practical matters generally.[16] Thus, as early as the meeting of 19 December 1660, the first after the completion of the Royal Society's organization, it was resolved "That Dr. Petty and Mr. [Christopher?] Wren be desired to consider the philosophy of shipping";[17] on 2 January 1660/61 Christopher Merret was requested to bring in a history of refining;[18] and a week later Evelyn was requested to bring in "his catalogue of trades."[19] Petty, Evelyn, Colwall, and Henshaw were all actively engaged in collecting material; some of this, like Henshaw's "History of the making of saltpeter" or Petty's "An apparatus to the history of the common practice of dyeing," were chosen in January 1664/65 for inclusion in Sprat's *History* of the Society.[20]

Hardly a meeting went by without some reference to trades, and the keener members provided surprisingly diverse materials for consideration. Thus on 8 January 1661/62 Evelyn brought in an account of a method for making marbled paper (tested a fortnight later), and in May he brought in "a written book of the history of the rolling-press," duly read from at a later meeting.[21] Not surprisingly, most of Evelyn's work of this kind was concerned with horticulture, culminating with his presentation of *Sylva* for licensing by the Royal Society.[22] Christopher Merret was another active participant: on 17 December 1662 he brought in an account of wine-making; and on 8 June 1664 he was made chairman of a committee for collecting histories of trades, in pursuance of which he brought in a catalogue of trades on 19 October 1664.

Merret must be remembered above all for producing, in 1662, an English edition of Antonio Neri's *L'Arte Vetraria*.[23] This is a supremely good example of the history of trades, for not only did Merret acknowledge Bacon's influence, but his work had twofold importance: it explored glassworking, with extensive comments on Neri's description of furnaces and processes and with enumeration of English practices, so that it is technologically informative; and it also contained a commentary on Neri's account of the chemical aspects of glassmaking, which is both an addition to the technology and of considerable scientific importance. Here is an excellent example of how the history of trades could be both practically useful and of lasting importance to an understanding of nature, far deeper than that provided by Petty or Evelyn. To be sure, Petty had been asked to investigate "the philosophy of shipping," but he was more inventive than philosophical, and while his "double-bottom" was a novel and interesting experiment in ship design—whose trials (in Ireland) the Royal Society followed attentively—Petty was not concerned with abstract principles; equally, Evelyn's interest was in horticulture, silviculture, and gardening, not in botany. Merret's work on glass has an altogether different character.

What else can one glean from the Society's recorded activities in its first four years? In part because of Boyle's interest in anything chemical, and Moray's in anything touching on mining, there was a great deal of chemical technology. Thus in 1661 we find Moray reading an account of the making of ceruse (white lead);[24] Colwall reading "an account of the making of green copperas" (green vitriol)[25] and another of alum;[26] Needham and Clayton reading on China varnish;[27] and Oldenburg asked to write an "account of the making of steel and lattin [brass] plates".[28] Then and later there were also several accounts of gunpowder, often at Prince Rupert's instigation.[29]

As time passed the interest became at once more miscellaneous and more encompassing. John Winthrop naturally spoke of American affairs, particularly as those related to ships.[30] The raising of oysters was discussed,[31] the manufacture of cider (extensively, partly because Evelyn was keen),[32] malting in Scotland,[33] a great deal about gardening—again because of Evelyn's interest and also that of Charles Howard (son of the Earl of Arundel and brother of the sixth duke of Norfolk), who contributed an account of saffron.[34] On 22 October 1662 Moray initiated a discussion on that topic of perennial practical interest, the making of iron with seacoal; there was a report on windmills in Holland,[35] which excited so much interest that John Pell promised to add to it; Moray reported on a mineral of Liège, a new French carriage, coal pits in Scotland, a new sort of rope used at Chatham, and on diving.[36] Tin mining was discussed on several occasions,[37] as was metal refining.[38] In 1664 there was a long series of reports and experiments on breaking wood by hanging weights upon it;[39]

this may have been partly practical but was largely experimental, and few conclusions seem to have been drawn from it. On 27 July 1664 the Royal Society Council greeted with joy the appointment of Hooke as "Professor of the histories of trades" with funding by Sir John Cutler,[40] while early in the new year there were accounts of the results of sending pendulum clocks to sea to measure longitude, and of the mercury mines in Friuli.[41]

This is all sufficiently miscellaneous, which is hardly surprising, for the meetings of the Royal Society, although orderly in themselves, tended to proceed by free association, with what was read or discussed by one member bringing varied ideas to the minds of others. And, as already remarked, the members differed radically in what they took to be the aim of the collecting of histories of trades; some were interested in fact, and utility, while others regarded the histories as merely tools to an end, that end being the understanding of nature through the advancement of natural science. Clearly, Oldenburg had been fully exposed to both aspects. It is notable, however, that very soon after he became established as, in effect, the corresponding secretary (after the second charter of 1663 granted the Society freedom to correspond without hindrance at home and abroad), he began to try to extract from his correspondents materials for what he called "a universal history of nature," which he maintained to be the Royal Society's principal aim. This very Baconian enterprise was of course never completed—probably it never could have been completed—but it extracted a great deal of information of both a scientific and a technological nature from England, Europe, the East and the West.

Not the least of Oldenburg's contributions to the Royal Society's reputation, and to its furthering an interest in technology, was his initiation of the *Philosophical Transactions*, the first European journal devoted exclusively to scientific matters and thus the oldest continuous scientific journal in the world. It is not always realized that the concept of such a journal was Oldenburg's alone. As already noted, the publishing of the *Philosophical Transactions* was Oldenburg's private venture; it was neither owned nor exclusively directed by the Royal Society, still less was it distributed by the Royal Society. To be sure,

each number of the *Philosophical Transactions* was licensed by
the Royal Society (it was a privilege granted to the Royal
Society in the first charter of 1662 that it might license books to
be printed by the Royal Society's printer), but in this the
journal did not differ from any work written by or edited by
any Fellow of the Royal Society. Except under very peculiar
circumstances (as during the Plague and immediately after the
Fire), all editorial responsibility was Oldenburg's, though he
might seek or accept help and advice from others.

There is no indication in his own correspondence about the
origin of this venture. The first reference to his plan that I
have come across is in a letter from Sir Robert Moray to
Christiaan Huygens of 3/13 February 1665, where he says
(apropos of Huygens's comments on the newly founded *Journal des Sçavans*):

Monsieur Oldenburg nous a fait voir un eschantillon d'un semblable dessein
bien plus philosophique, et nous faisons estat de l'y engager, sil se peut faire.
Il ne se meslera pas des choses Juridiques ny Theologiques [as the *Journal des
Sçavans* was pledged to do], mais outre les choses philosophiques qui nous
viennent de delà la mer il publiera les experiences, au moins les chefs, qui se
font icy, mais ce ne sera quune fois le mois, en Anglais, et une fois en trois
mois, en latin.[42]

Needless to say, Oldenburg never himself issued any numbers
in Latin, although others were to translate the first four volumes. A month later, at the Royal Society's Council meeting
on 1/10 March 1664/65,

It was ordered . . . that the *Philosophical Transactions*, to be composed by Mr.
Oldenburg, be printed the first Monday of every month, if he have sufficient
matter for it, and that the tract be licensed by the Council of the Society,
being first reviewed by some of the members of the same; and that the
president be desired now to license the first papers thereof, being written in
four sheets of folio, to be printed by John Martyn and James Allestry,
printers to the Society.

The first number is duly dated "Munday, March 6. 1664/5."
Although in his introduction Oldenburg speaks only of "the
improvement of Philosophical Matters," he by no means intended to exclude technological ones either, and this first
number contains "An accompt of the improvement of Optick
Glasses," a brief summary "Of a Peculiar Lead-Ore of Ger-

many, and the Use thereof," a note "Of an Hungarian Bolus, of the same Effect with the Bolus Armenus," a relation "Of the New American Whale-fishing about the Bermudas," and "A Narrative concerning the success of Pendulum Watches at Sea for the Longitudes"; so of ten items, no fewer than five might count as technological. The first item, on "optick glasses," was clearly of specifically scientific importance; the lead ore was of chemical interest, and the bolus of medical. The account of whale fishing was bound to interest those members of the Society who on 4 November 1663 had listened to the account of whale fishing in Greenland by Mr. Gray of the Greenland Company. And the account of pendulum watches at sea had been presented in essence by Moray on 11 January 1664/65; its importance for the attempt to establish a reliable method of determining longitude at sea hardly needs emphasizing.

Later numbers of the first volume contained items as diverse as an account of the proposed Languedoc canal; Walter Pope's relation of the mercury mines at Friuli (already read to the Royal Society on 11 January), which contains technical description, economic analysis, and a description of mercury poisoning among the miners; Moray's account of Liège minerals (read to the Society on 10 June 1663), of mine damps in Scotland (a subject of perennial interest), and more about mining in Liège, especially on the ventilation of the mines; much more on telescope lenses and on burning mirrors; a good deal on agriculture, dyes, navigational problems, and the cultivation of silkworms; and a rumor about the making of porcelain. The whole seems almost as diverse as the matters discussed in the Royal Society, differing at first only in being more international; but gradually the journal came to have a greater consistency, especially as the items grew fewer but longer.

Almost as interesting as what Oldenburg put into the *Transactions* is what he left out. There is, for one thing, no mention of Petty's double-bottom ship that occupied so much of the attention of the Royal Society during 1664 and 1665. In fact, it had been intended that the third number of the *Transactions* should have a detailed account, but (on 26 April 1665) the council "resolved that the publication should be deferred till the King had been made acquainted with the particulars relating to the said ship," and it was later (3 May) decided that it should be deferred "to another time." This other time never came, possibly for the same reason that led the council two years before to refuse to take an official view[43]—namely, that "the matter of navigation being a state concern was not proper to be managed by the Society"—but probably also because the trials of the ship were too inconclusive and any comment might seem excessively controversial.

Certainly in 1673 (Vol. VIII) Oldenburg showed no hesitation in publishing a letter on the advantages to the navy and to shipping generally of sheathing ships in lead to prevent the incursions of the teredo worm in tropical waters, very much a state concern; and there are reviews of several works on navigation, naval architecture, and similar matters. It must be added that in Volume I Oldenburg also ignored various items of interest discussed by the Society, such as Prince Rupert's recipe for gunpowde · (given him by "FMZB")[44] and his water-raising engine. The other topic extensively discussed by the Society in 1665 which Oldenburg never touched upon in the *Transactions* was the design of new carriages and chariots, a topic of perennial interest in both London and Paris. Oldenburg had come across this topic during his time in France; but though it occurs as an item of interest in letters from that time onward, he apparently did not regard it as a subject appropriate to his journal.

Having finished two years' editing (Vol. I contains twenty-two issues), made difficult by the Plague year and the Fire of London (which forced the publication of several numbers at Oxford, with Boyle and Wallis as assistant editors), Oldenburg could look back with some complacency on his accomplishment, and fairly so, remarking:

I think I may safely assume, that in these Fragments, something hath been contributed to sowe such seeds, as may somewhat conduce to the illustration and improvement of philosophy, and of all laudable and useful Arts and Practices.

And again:

As for the Growth of Arts and Inventions, I think it may be justly said, That these our Entries sometimes assist and promote their Improvements. And the same will hereafter remain faithful Records to shew, By what Essays,

Emulations, and Encouragements these Noble Arts advanced to perfection. And a punctual information of these Gradual Processes, may be instructive to promote other Inventions.[45]

Pursuant to which, this number saw several articles relating to the compass needle and to navigation. Subsequent issues saw descriptions of grafting and maintaining fruit trees, accounts of Richard Towneley's instrument for "dividing a foot into many thousand parts" (a micrometer, for astronomical purposes),[46] "Experiments for Improving the Art of Gunnery,"[47] queries on mines, and more about whale fishing.

Let us consider more generally the topics Oldenburg touched upon in his dozen or so years of editing the *Philosophical Transactions*, always bearing in mind that he was reviewing the work of others, and could only go as far as his sources— letters, other journals, papers read to the Royal Society, and published books—served him. Navigation, already mentioned, was a continuing interest of all right-thinking Englishmen in the seventeenth century, and it is no surprise to find, year by year, items on longitude determination, determination of magnetic variations, improvements of compass needles, lead sheathing of ships, naval architecture, and accounts of voyages.

Similarly with mining. The Royal Society had an early interest in foreign mines, an interest especially catered to by Edward Browne during his Carpathian journeys of 1669 and 1670. Oldenburg was apparently always on the lookout for information; and about English mines there are discussions of the effects and behavior of the destructive mine damps, as well as more general accounts. (Incidentally, Oldenburg helped to preserve a large vocabulary of mining terms, which enormously puzzled the German who translated the first four volumes into Latin.)

A subject that occurs in most volumes is agriculture, broadly interpreted: planting, grafting, and cultivation of fruit trees (in this one sees the influence of Evelyn and of John Beale); cultivation of melons (a delicate tribute to Charles Howard, who was particularly interested in this topic); the making of cider and related beverages (not only owing to Evelyn, but also to Beale who came from Herefordshire and had, as the *Dictionary of National Biography* quaintly puts it, "an heridatary interest in cider"). There are frequent accounts of husbandry and reviews of books on agriculture, horticulture, sheep farming, and related subjects. It must be noted that Beale kept a vigilant eye on the *Philosophical Transactions*, writing Oldenburg immensely long, nearly illegible letters intended for publication—often commenting on previous numbers, praising Oldenburg for his attention to such matters, and scolding him for slighting them or for extolling foreign practices—acting rather as a Hartlibian critic.

In the long run, however, Oldenburg alone was responsible for choosing the contents, and he was often more perceptive and more internationally oriented than his colleagues. Thus he inserted no fewer than three quite long accounts of the Languedoc canal: one, already mentioned, was a summary of Petit's proposals in 1665; the second and third, in 1670 and 1672, were translations from the French of printed narrations.[48] (The second account earned him a rebuke from the xenophobic Beale, who thought that publicizing the ingenuity of foreigners would cause despondency among the English, then at war with the French.)[49]

Oldenburg persisted in publishing numerous accounts of foreign "trades," especially French accounts gathered from his increasingly extensive correspondence: notably, accounts of the Orléanais method of making vinegar and of the methods used at La Rochelle for the manufacture of sea salt.[50] It was perhaps in consequence of the reading of this latter account at a meeting of the Society that inquiries about the saltworks in Cheshire were sent to Dr. William Jackson of Nantwich; the replies were in turn printed in extenso in the *Transactions* on two separate occasions in 1669.[51] The next year Oldenburg printed an account of the mining of salt in Poland, sent to Hooke some years previously, and at the very end of the year he was able to print an account of the discovery of rock salt in Cheshire, sent by a nonconformist preacher, Adam Martindale, in letters to Oldenburg.[52] These, except for the last, are true histories of trades as the Royal Society originally conceived them when the Fellows agreed to undertake the compiling of a wide diversity of them in 1664;[53] but on the whole they had tended to lose interest in the histories as the years went by. Although a number were brought in, most remained in the register book, whence some were dredged up by Nehemiah Grew in 1678 when he needed material to complete Volume XII of the *Philosophical Transactions* (see nos. 137, 138, and 142) after Oldenburg's death.

Very possibly Oldenburg was less interested than some other members of the Royal Society (notably Hooke, whose attitude never changed); equally possibly those who brought in the accounts were not very keen to have them printed. Certainly Oldenburg was ready to print what Evelyn and Beale urged on him in regard to agriculture; and he readily printed anything novel. For example, in 1674 Charles Howard wrote "Brief Directions how to Tanne Leather according to a New Invention," which Oldenburg printed in extenso[54] with a careful description of the "engine" used to cut up the wood that was to be employed instead of bark; here it was carefully noted that the new invention had been "experimented and approved by divers of the principal Tanners using Leaden-Hall Market." Perhaps Oldenburg was in part thinking of this contribution when, in the introduction to Volume X (1675),[55] he wrote, "The Ingenuous Arts do furnish Employments for the younger Descendents of generous Families," for the Howards were a very great family indeed.

One subject that obviously appealed to Oldenburg enormously, as indeed it did to the inventive minds of the age, was the invention of instruments. By no means every such invention was noted in the *Philosophical Transactions*, but it is amazing to see how many were described there in the dozen years during which Oldenburg edited the journal. The very first number contained as its first item "An accompt of the improvement of Optick Glasses" (by Campani), and the last of Oldenburg's *Transactions* (no. 136, 25 June 1677) contains an account "of a very useful and cheap Pump, contrived by . . . Mr. Conyers, the Author of the Hygroscope described in number 129." (In this case, the pump was a large one, which was tried during the enlarging of the "New Canal of Fleet-River in London," but a pump for scientific work would have received the same publicity.) In between came a host of inventions: a water bellows, leather tubes to improve the sight, Wren's perspective-drawing device, a diving bell, Samuel Morland's loudspeaking trumpet (a kind of megaphone), drawing instruments, Morland's pump, Boyle's hydrostatical balance, a hydraulic engine from France, hygroscopes, and assaying devices.

Two kinds of inventions interested Oldenburg especially: optical devices and timekeeping devices. Accounts of telescope lenses, burning mirrors, attempts at parabolic lenses, microscope lenses, machines for grinding and polishing lenses, Newton's reflecting telescope, and Cassegrain's reflector all appeared in due season, and Oldenburg rightly perceived the importance of each. Oldenburg also gave due, but not excessive, consideration to a related *scientific* problem: the use of telescopic sights on astronomical instruments, as expressed in a long and extremely vituperative disagreement between Hevelius (who clung to open sights) and Hooke (who rightly but not very tactfully or adroitly) claimed vast superiority for telescopic sights. Hevelius was of course the far more experienced observer, and as John Wallis commented privately to Oldenburg, was almost certainly seeing more with the naked eye, and with more accuracy, than most astronomers could do with telescopic sights. The quarrel raged for over a year, even embroiling Flamsteed, to his own astonishment (for although he agreed with Hooke about the virtue of telescopic sights he never overtly said so to Hevelius). Oldenburg managed throughout to keep Hevelius's friendship and to publish as little of the quarrel as he could.[56]

He was less successful in managing the publicity he rightly thought important in regard to timekeeping devices, though he began well with "A Narrative concerning the success of Pendulum-Watches, at Sea for Longitudes" in the first number of the *Transactions*. This—though it describes Robert Holmes's voyage to Guinea in 1664 carrying watches, as the account says, "first Invented by the Excellent Mounsieur Christian Hugens of Zulichem, and fitted to go at sea, by the Right Honourable, The Earl of Kincardin, both Fellows of the Royal Society"—had already been published in the *Journal des Sçavans* for the previous month. Sir Robert Moray had sent the "Narrative" to Huygens, and he in turn had sent it to Jean Chapelain, who had both the "Narrative" and Huygens's comments on it printed in French.[57] All Oldenburg did was translate the whole into English to provide an account of a highly successful example of international cooperation.

Huygens, clearly pleased, gladly acceded four years later to Oldenburg's request for permission to print more.[58] As Oldenburg put it in his letter, "let me inform you that our common friends here think well of having printed here, under your name, your instructions for the use of pendulum clocks at sea, translated into English [from Dutch], and somewhat enlarged and illustrated by Lord Brouncker and Sir Robert Moray; it could be extremely useful to the public." Huygens agreed immediately, remarking: "If my consent were required to print my instructions for clocks I would gladly give it. But as it is already public property because of the edition I had printed in Dutch, I no longer have any rights on it." And he went on to welcome Brouncker's and Moray's modifications without seeing them, only specifying that it should be made clear that these modifications were not his. (In the event, Brouncker and Moray became simply "two other Eminent Members of the Royal Society" in the printed account.) Within a month, the whole appeared at length in the *Philosophical Transactions*.[59] This all must have been at the instigation of Brouncker (who, it must be remembered, was a member of the Navy Board) and Moray, for there is no hint of its being discussed by the Royal Society at large.

At this time (1669–1670) Oldenburg printed nothing of the discussions actually going on at the Society's meetings about Hooke's design for a "magnetical watch." This must not be construed as prejudice on Oldenburg's part. He and Hooke, the two paid officers of the Royal Society, were working together quite harmoniously. It is probable that the reason was the inchoate state of Hooke's design and his own reluctance to have his inventions discussed outside the Royal Society's meetings; for Oldenburg seems always to have been ready to publish inventions when the author gave permission, whereas Hooke was often reluctant to record what he was willing to say spontaneously—witness the number of his papers and lectures never registered (though the Royal Society asked that they be registered) and, especially later, the number of occasions when the manuscript minutes and their drafts reveal that Hooke was asked to write out his spoken comments and failed to do so, with the result that it is only recorded that he spoke at a meeting, not what he said.

It was indeed Hooke's curious reluctance to have what he

said made public combined with his intense jealousy for the priority of the ideas that sprang to his highly creative mind which was to cause the disastrous rift between him and Oldenburg in 1675. Volume X of the *Philosophical Transactions* opened with (in no. 112, 25 March) "An Extract of the French Journal des Sçavans [of 25 February, n.s.], concerning a New Invention of Monsieur Christian Hugens de Zulichem, of very exact and portative Watches."[60] This is a brief but exact description, with a figure, of Huygens's spring balance as applied to watches, an invention he had made and put into practice the previous January and described to Oldenburg in a letter dated February 1674/75, a letter read to the Royal Society on 18 February.[61]

About this occasion Hooke wrote in his *Diary*, "Zulichems spring watch spoken of by his letter. I shewd when [where ?] it was printed in Dr. Spratts book. The Society inclind to favour Zulichems." The entry in Sprat's *History of The Royal Society* mentions[62] "Several new kinds of Pendulum Watches for the Pocket, wherein the motion is regulated, by Springs, or Weights, or Loadstones, or Flies moving very exactly regular," which is hardly the same as having a functioning spring-balance watch already constructed, as Hooke was to do only *after* learning of Huygens's success. It was undoubtedly in an attempt to avoid personalities that Oldenburg printed, not Huygens's letter to himself, but a translation of what had already been printed abroad. And in the next number he did the same for Leibniz, printing an English translation of his very strange and impractical form of spring balance as published in the *Journal des Sçavans*.[63] Hooke remained convinced that he had been the first to think of such a balance and that, what is more, his invention as described by him to the Royal Society had been "betrayed" to Huygens, a foreigner, by Oldenburg, another foreigner. (He conveniently forgot that Huygens was also a Fellow of the Royal Society and that it was part of the secretary's duty to send out news of interesting developments presented at meetings.) He was assured of the betrayal by the fact that Huygens offered English patent rights to either the Royal Society or Oldenburg, and Brouncker, as president, encouraged Oldenburg to claim them. Hooke promptly set to work to make a spring-balance watch for the king to offset Brouncker's demonstration of Huygens's successful invention. Hooke did make a satisfactory watch, which in turn made the king decide not to grant a patent for Huygens's version; but the publication of Huygens's invention in the *Translations* still rankled. Preparing some of his Cutlerian lectures on instruments for publication in late September, Hooke wrote a diatribe against both Oldenburg and Huygens. *A Description of Helioscopes and some other Instruments* was printed early in October[64] with a "Postscript" attacking Oldenburg for publishing an account by a foreigner of what an Englishman had invented and published long before. In it Hooke insisted on his own priority and that Huygens, being aware of his ideas, had no claim to be the inventor. Further, he denounced Oldenburg, who "was not ignorant" of all these things.

Oldenburg was not a touchy man normally, but the accusation of underhandedness and partiality moved him deeply. In his very next issue (no. 118, 25 October), instead of the usual review of Hooke's *Helioscopes*, he gave the title and promptly turned to the postscript. He clearly thought the best defense was an historical narrative, and so he gave one, stating carefully, "Tis certain then, that the Describer of the Helioscope, some years ago, caused to be actually made some Watches of this kind, yet without publishing to the world a Description of it *in print*; but it is as certain, that none of those Watches succeeded, nor that any thing was done since to mend the Invention, and to render it useful, that we know of, until Monsieur Hugens, who is also a Member of the Royal Society . . . sent hither a letter,"—the letter claiming to have made the invention, which he then described in detail. Oldenburg went on to explain how the *Philosophical Transactions* item came to be printed, adding, "whether there be any thing . . . that deserves the name of *unhandsome proceedings* [Hooke's term], he very willingly leaves to all Ingenuous Readers to judge," and to defend his practice of publishing new inventions and to note that, when he had showed Hooke Leibniz's invention, Hooke had not claimed that Leibniz had stolen it from him.[65]

If Oldenburg hoped that Hooke would stop here he was mistaken. Hooke's resentment remained high, and by the end of the summer he had ready for the press another collection of inventions to which he could add a postscript attacking Oldenburg's defense. *Lampas* was licensed on 21 August 1676, and not by the Royal Society as would have been proper since the printer was as usual John Martyn, "Printer to the Royal Society." The book was finished a month afterward.

In his Postscript to *Lampas* Hooke attacked Oldenburg principally on two points: first, that Oldenburg had said that none of his watches had "succeeded" in the past, and second, that he had not published any account of them. To the first point Hooke asked, "how could he be sure of a Negative? Whom I have not acquainted with my Inventions, since I looked on him as one that made a trade of Intelligence." (To which one might reply that if this were so Oldenburg could not have "betrayed" him to Huygens.) To the second point he answered that he had spoken of the watches in his public lectures, "made and shown to thousands both English and Foreiners, and writ of to several persons absent," as well as "published to the world" in Sprat's *History*. (And if this were so, one might again ask how Oldenburg could have "betrayed" him.) The rest of his attack is a confused, irrelevant, and personal grumble.

Oldenburg was clearly touched in his professional honor. *Philosophical Transactions* number 128 (25 September 1676) was licensed at the first council meeting after the summer vacation, on 3 October.[66] Oldenburg had just time to insert an "Advertisement To intimate, that the Publisher of this Tract intends to take another opportunity of Justifying himself against the Aspersions and Calumnies of an immoral Postscript put to a Book called *Lampas*, publisht by Robert Hooke: till which time, 'tis hoped, the Candid Reader will suspend his Judgment." Oldenburg lost no time in appealing to the Council of the Royal Society, and on 12 October "It was ordered, that the Society's printer, Mr. Martyn, he required to give notice in the *Philosophical Transactions* next to be printed, of what the Council was informed he had declared, viz., that the tract called *Lampas* . . . was printed without the leave or knowledge of the Council of the Royal Society, and that the said printer had seen nothing of the postscript thereof before it was printed off, nor knew any ground for the aspersions contained therein" under threat of being replaced as the Royal

Society's printer. It was not until three weeks later, however, on 2 November, that the council appointed Dr. Croune and Mr. Hill to prepare such a statement, at the same time authorizing Oldenburg to print part of Huygens's letter to him which spoke of the assignation of patent rights. Nor was it until 20 November that Croune and Hill brought in a brief statement exonerating the council from knowledge of Hooke's attack and stating that Oldenburg "hath carried himself faithfully and honestly in the management of the Intelligence of the Royal Society, and given no just cause of such Reflections." This, with the authorized extract of Huygens's letter, is printed at the end of *Philosophical Transactions* number 129, dated 20 November 1676;[67] no doubt Oldenburg held back the printing of this number (there was no *Phil. Trans.* for October) until he could insert the justification.

Hooke was not a member of the Council, but he was kept informed by those who were, and after the Council meeting of 12 October he thought of resigning—and as quickly forgot about it, only noting in his diary for 20 October, "Grubendolian Councell order" (but there is no record of a Council meeting that week). Of the next meeting (on 2 November) Hooke noted, "Grubendolian Councell . . . Croon a dog." Yet he could write calmly of ordinary Royal Society business, and of Croune and Hill, with both of whom he was normally friendly. Of the Council meeting of 20 November Hooke wrote only, "packed Councell," as of the elections on 30 November he wrote, "Much fowl play," because he reckoned he had received fifteen votes to be a member of the Council, while Oldenburg counted only thirteen votes (it is not clear how Hooke thought he knew how many votes he had received).

Those who know Hooke will not be surprised that he remained friendly with Croune and Hill; but the old easy relations with Oldenburg were gone forever. His name is mentioned from time to time in Hooke's *Diary*, and they met in the way of business; but their previous intimacy was ended forever, to be replaced by resentment, which flared up again at Oldenburg's death although Hooke refrained from any more "postscripts." What Oldenburg felt after November 1676 we do not know; the depth of his first resentment is displayed in his own drafts for the reply to *Lampas*,[68] but he presumably accepted the Royal Society's decision to print only the simple disclaimer of 20 November.

This account of such an undignified squabble may seem a digression, but it is a necessary adjunct to any consideration of the publishing of technological accounts in the *Philosophical Transactions* for the light it throws on seventeenth-century attitudes to technological innovation. Later centuries have, of course, seen other examples of priority disputes, and the phenomenon of simultaneous invention has been much investigated by historians. Nor are xenophobia and chauvinism unique to the seventeenth century. Leaving personalities aside, one can see that the case is far from simple. Hooke had certainly thought of the possibility of applying a spring balance to a watch, Sprat had mentioned the idea in his *History*, and Huygens probably knew this; but Hooke, in spite of his claim, never made a going spring-balance watch before he read Huygens's account of his own such watch. Very probably Hooke saw no difference between having the idea for such a watch, and actually making it. His was an inventive mind, and when he did try he was able to improve upon Huygens's first design. No doubt too he was right to see that Brouncker and Oldenburg were partisans of Huygens.

Historians have often shared Hooke's inability to perceive the difference between having an idea for a technological process and making it work or putting it into production. There *is* a difference, however, and Oldenburg was surely right to see Huygens as the true inventor. And even if he were wrong, it is hard to see how he erred in printing in English the published account from the *Journal des Sçavans*—nor could it have been wrong for the Council to approve its inclusion. The whole episode demonstrates the difficulties involved in the publication of technological discovery and invention. How could priority be established without publication? But could not publication lead to development of one's ideas by others, including foreigners? Even Hooke found himself publishing scientific discoveries made by foreigners (in *Cometa*, before he began the *Philosophical Collections*).[69] There was an unresolvable dilemma here, made more difficult by the fact that some at least of these foreigners were also Fellows of the Royal

Society (Huygens, most notably, and Hevelius, Cassini, Leeuwenhoek, and Malpighi).

Yet in the months following his great quarrel with Hooke,
Oldenburg did continue to publish letters from foreigners in
the *Philosophical Transactions* (mainly on astronomical subjects,
but also on various kinds of phosphorus), and he did continue
to publish on agriculture and inventions (a hydraulic engine
and telescope lenses—from the *Journal des Sçavans* once
again—and Conyers's hygroscopes and pump). His last issue
of the *Philosophical Transactions* was that for June 1677. It
continued the practice of previous issues by publishing a mixture of science and technology. And if the science—Roemer's
measurement of the velocity of light, and Leeuwenhoek's
microscopical discoveries—is of more lasting importance than
the technology—descriptions of Conyers's pump, a celestial
globe made by a Paris watchmaker, diamond mines, by "The
Earl Marshal of England" (Henry Howard)—does this not
hint at one reason the Royal Society was not, in 1677, pursuing
histories of trade as avidly as it had in 1664? Clearly, the
Fellows had not lost interest in technological processes; but
possibly they were coming to realize that this interest was less
immediately profitable "for improving natural knowledge"
than they had originally hoped.

Oldenburg had established the *Philosophical Transactions* on
such a firm basis that the Royal Society was to find it a necessity.
And his balance between science and technology was retained
throughout the seventeenth century. Perhaps under Hooke's
influence, many old accounts of histories of trades, predating
1665, went to make up a large share of the remainder of
Volume XII, of the *Philosophical Transactions* published after
Oldenburg's death, though Hooke's *Philosophical Collections*
and later numbers of *Philosophical Transactions* returned to the
balance favored by Oldenburg and were entirely up-to-date.

It is in no way surprising that before the end of the century
an account of Savery's engine was published in the *Philosophical Transactions*. That there is no such account of Newcomen's
perhaps reveals something of the changing nature of both
science and technology in the eighteenth century, their growing separation, and the faint beginnings of a new relation
between them.

NOTES

1. *Journal of the History of Ideas*, II (1941): 33−60, a study as fresh and thought-provoking now as when it was written. I am, and have long been, much indebted to Houghton's brilliant, perceptive, and subtle analysis.

2. See the "Advertisement" at the end of no. 12 (7 May 1666), 213−214, and his correspondence with the translator C. Sand in *The Correspondence of Henry Oldenburg*, ed. A. R. and M. B. Hall (Madison: University of Wisconsin Press, 1973), IX, passim, esp. 517 n. 3.

3. For a summary of his early life, with sources, see A. R. and M. B. Hall, "Some Hitherto Unknown Facts about the Private Career of Henry Oldenburg" and "Further Notes on Henry Oldenburg," *Notes and Records of the Royal Society 18* (1963): 94−103 and *23* (1968): 33−42, respectively.

4. *Correspondence* (Madison: University of Wisconsin Press, 1965), I: 90−91, letter 39 of April 1656.

5. He kept a notebook in which he copied important letters sent or received and extracts of the more informal ones. This is preserved in the Royal Society as MS MM 1.

6. For all this, see *Correspondence*, I, passim.

7. See Charles Webster, *The Great Instauration* (London, Duckworth 1975), esp. Chap. V.

8. *Correspondence*, I: 102.

9. As, for example, "The Optical lantern," for which see ibid., I: 201.

10. Ibid., 270.

11. See the copies and extracts sent to Evelyn in British Museum Sloane MS 15948, fols. 71−74, 76, to be printed in the final volume of the Oldenburg *Correspondence*. I have to thank Dr. Robert Frank for calling my attention to these MSS.

12. *Correspondence*, I: 406.

13. For all this see Thomas Birch, *History of the Royal Society*, 4 vols. (London, 1756−57, New York: Johnson Reprint, 1968), I:3−4, 8.

14. Ibid., pp. 15, 23, respectively.

15. Ibid. p. 5.

16. Besides the article of Houghton (see note 1 above), this topic has been extensively considered by Robert K. Merton in "Science, Technology and Society in XVIIth Century England," *Osiris* 4 (1938): 360−632; reprinted with a new preface in paperback form (New York; Harper Torchbooks, 1970). Merton was much less careful in his analysis than one would wish, and he was inclined to assume technological influences on somewhat flimsy evidence. His main interest was in the effect of utilitarian interests on scientists, but it is nonetheless a vitally important work.

17. Birch, *History*, I: 7. Petty was known to have a deep interest in navigation and ship design. Future quotations from this source will not be referenced when the date of the meeting is given.

18. This was to be printed in *Phil. Trans.* XII, no. 142 (1678/9): 1046. Merret, although best known as a botanist, was also a chemist, as well as a physician.

19. Evelyn had been working at this at least since 1656; in 1703 he

remembered that Boyle, whom he had met early in that year, had encouraged him in this, to which he had been incited by Hartlib. See Houghton, "History of Trades," p. 46.

20. Read on 14 August 1661 and printed in Thomas Sprat, *History of the Royal Society* (London, 1667), p. 260, and read on 7 May 1662 and printed in Sprat, p. 284, respectively. For their consideration, see Birch, *History* under 11 January 1664/5 and 18 January 1664/5. The final printing of Sprat's book was of course delayed by the Plague and the Great Fire of London. Petty was, it should be remembered, the son of a clothier, and he brought in a "History of Clothing" on 27 November 1661.

21. On 14 May 1662 and 18 June 1662, respectively.

22. The order for its printing (after perusal by Wilkins, Goddard, and Merret) was given on 18 March 1662/3.

23. Published at Florence in 1612; Neri described himself as a priest, so his work may be regarded as part of the Renaissance intellectual's interest in technological practices; he clearly had frequented glassworks. Merret's English edition was published at London, under the title *The Art of Glass*; there was a Latin edition, which was published at London in 1668.

24. Written apparently by Sir Philiberto Vernatti, it was read on 13 March 1660/61.

25. On 3 April, Colwall, Boyle, and Merret had been "desired, to draw up an account of the manner of making vitriol"; Colwall's account was read on 8 May and ultimately printed in *Phil. Trans.* XII, no. 142 (1678); 1056–1059.

26. On 14 August, printed in *Phil. Trans.* XII, no. 142 (1678): 1052–1056.

27. On 30 October and 20 November, respectively; their accounts were preserved in the register book, as Birch, *History* carefully noted.

28. On 30 October; he is not recorded as having done so.

29. Besides Henshaw's investigation of nitre (see note 20 above) and Schroter's paper on saltpetre (7 January 1662/3, printed in Birch, *History*), there were a German account of gunpowder (8 January 1660/61), a letter from Zurich read on 8 October 1662, others sent by Prince Rupert on 22 July 1663 and 2 December 1663, Hooke's "powder-tryer" discussed on 9 December 1663, and an experiment on the force of gunpowder, 18 May 1664.

30. On 16 July 1662 "of the manner of making Tar and Pitch in New England" and on 24 September 1662 on shipbuilding; both these papers were preserved in the register book.

31. By Colwall on 29 October 1662 and by Haak on 19 November 1662; neither paper was registered.

32. Dr. Smith of Gloucester sent in a paper read on 26 November 1662, while a paper of Beale's was read on 10 December 1662 (it is printed in Birch, *History*, I:144–152), and his letter to Oldenburg of 21 December 1662 (*Correspondence*, I: 481–483) was read on 7 January 1662/3, after which Evelyn was encouraged to work on what became *Pomona* (London, 1664).

33. By Moray on 7 January 1662/3; it is printed in Birch, *History*, I: 169–71.

34. Read first on 2 December 1663 and, with additions, again on 9

December 1663. It was to be printed in *Phil. Trans.* XII, no. 138 (25 March 1678): 945–949.

35. By "Mr. Bruce" (i.e., Alexander Bruce, about to become the Earl of Kincardine); it was registered.

36. On 10 June 1663, 11 November 1663, 2 December 1663, 16 March 1663/4, and 18 May 1664, respectively.

37. On 25 May 1664, Dr. Edward Cotton brought in an account (registered, and printed in Birch, *History*, I: 429–430); there is a more detailed account of tin mines in Cornwall by Merret printed in *Phil. Trans.* XII, no. 138 (25 March 1678): 949–952.

38. See 29 June 1664, 6 July 1664 and Goddard's account of "Refining Gold with Antimony," *Phil. Trans.* XII, no. 138, 953–961.

39. This investigation was begun on 23 March 1663/4 and continued for several months; as a form of inquiry into the strength of materials, it may have been influenced by someone's reading of Galileo, although just possibly it related to an interest in shipbuilding.

40. For Hooke's own account of the founding of the Cutlerian lectures, see Hooke's letter to Boyle of 6 October 1664 (Thomas Birch, *The Works of the Honourable Robert Boyle*, 5 vols. [1744], V: 537, extracted in Birch, *History*, I: 473 note). The Cutlerian Lectures were reprinted by R. T. Gunther in *Early Science in Oxford* (Oxford, 1931), Vol. VIII; although there are accounts of many scientific instruments invented by Hooke, and of lamps and clocks, there is nothing here specifically on histories of trades.

41. Reported on 11 January 1664/5; see below.

42. *Oeuvres Complètes de Christiaan Huygens* (La Haye, 1893), V: 234. In translation: "Mr. Oldenburg has shown us a sample of similar design, but much more philosophical [i.e., scientific], and assures us that he will undertake it, if he can. He will not meddle with legal or theological matters [as the *Journal des Sçavans* did], but besides the philosophical news which reaches us from overseas, he will publish the experiments, at least the principal ones, which are made here; but this will only be once a month in English and once every three months in Latin."

43. Birch, *History*, I: 249.

44. Ibid., I: 281, 286.

45. Preface to Vol. II, no. 23 (11 March 1666/7): 409 and 411.

46. *Phil. Trans.* II, no. 25 (6 May 1667): 457–458, followed by Hooke's claim to have invented other ways of doing the same thing, printed in Vol. II, no. 29 (11 November 1667): 542–544.

47. Vol. II, no. 26 (3 June 1667): 473–477.

48. In Vol. I, no. 3 (8 May 1665); 41–43; in Vol. IV, no. 56 (17 February 1669/70): 1123–1128, with map; and in Vol. VII, no. 84 (17 June 1672); 4080–4086.

49. *Correspondence* (Madison: University of Wisconsin Press, 1969), VI: 561.

50. Sent by one Elie Richard, a physician in those parts; see *Correspondence*, VI: 82, 200, and *Phil. Trans.* V, no. 61 (18 July 1670) and IV, no. 51 (20 September 1669): 1026–1028.

51. In Vol. IV, no. 53 (15 November 1669): 1060–1067, with figures, and Vol. IV, no. 54 (13 December 1669): 1072–1079.

52. In Vol. V, no. 61 (18 July 1670): 1099–2002 and Vol. V, no. 66 (12 December 1670): 2015–2017, respectively.

53. On 19 October 1664; see Birch, *History*, I: 476.

54. In Vol. IX, no. 105 (20 July 1674): 93–96; there is a separate printed sheet (on one side only) preserved in Royal Society Classified Papers, XXV, no. 94.

55. Vol. X, no. 112 (25 March 1675): 255.

56. This controversy is best seen in the letters exchanged among Hevelius, Oldenburg, and Flamsteed; the first appear in *Correspondence*, XI (London: Mansell, 1977), and the rest will appear in *Correspondence*, XII and XIII (London, Taylor & Francis, 1983).

57. See Huygens, *Oeuvres Complètes*, VII: 204–206 and 223. For the report to the Royal Society, see note 41 above.

58. See *Correspondence* (1968) V: 503 and 557, respectively (also printed in Huygens, *Oeuvres Complètes*).

59. Vol. IV., no. 47 (10 May 1669): 937–953.

60. P. 272–273.

61. *Correspondence* (Mansell, 1675), X:184–187.

62. P. 247, where moreover Hooke's name is not mentioned.

63. *Phil. Trans.*, X, no. 113 (26 April 1675): 285–288 from the *Journal des Sçavans* of 25 March 1675, n.s.

64. According to *The Diary of Robert Hooke 1672–80*, ed. H. W. Robinson and Walter Adams (London: Taylor and Francis, 1935), it was available by 11 October. Hooke's side of the quarrel can be followed in his *Diary*; Oldenburg's in *Correspondence*, XI and XII.

65. Vol. X: 440–442.

66. See Birch, *History*, III: 319 and the imprimatur; the original is British Museum Add. MS 4441, fol. 60. For the "Advertisement," see *Phil. Trans.* XI: 710.

67. *Phil. Trans.* XI: 749–750. The original drafts are in British Museum Add. MS. 4441, fols. 61–63.

68. Preserved in British Museum Add. MS. 4441, ff. 59 and 100.

69. *Cometa* (London, 1678) contains several excerpts from the *Journal des Sçavans* and two letters by Leeuwenhoek. Hooke's *Philosophical Collections*, a replacement for *Phil. Trans.*, began publication in 1680.

III

THE BIRTH OF THE MODERN
SCIENTIFIC INSTRUMENT, 1550–1700

Albert Van Helden

The word *instrument* has a variety of meanings, all clustered around the notion of tool or device.[1] When we modify it to *scientific instrument*, we mean a device used by scientists to investigate nature qualitatively or quantitatively. We tacitly assume that: (1) there is a proper, even essential, place for such devices in the study of nature since the human senses alone are too limited for most scientific investigations; (2) the results or readings obtained with them are usually beyond question;[2] (3) scientific instruments are based on undisputed scientific principles, such as the law of the lever; (4) newer instruments are more accurate, powerful, or convenient than older ones, the limiting factor usually being the state of the art in contemporary technology. Looking back at the period 1550–1700, which saw the birth of modern scientific instruments, we find, however, that these modern assumptions concerning instruments were not present and that these views became "truths" only as the use of instruments became commonplace.

Among the new instruments that were brought to bear on nature during this period were precision devices for measuring the positions of heavenly bodies, the telescope, the micro-

49

scope, the thermometer, the barometer, the air pump, and the pendulum clock. This list is by no means exhaustive. I have omitted the balance because, although it had been in use for a long time, it did not become an important scientific instrument in chemistry until the eighteenth century.[3] The list should perhaps also include such things as lenses, prisms, and calculating machines; but for the purposes of this discussion it is sufficiently illustrative. In order to discuss these instruments and draw some general conclusions, I shall begin by considering the status of qualitative observation and precision measurement and then discuss their development in the context of the two new methods of science, mathematical and experimental, as they gained momentum in the seventeenth century.

OBSERVATION

This method of science was used in antiquity and is particularly associated with Aristotle; it is still practiced in many branches of science, especially the life sciences. Nature is observed with the unaided senses, and only the simplest tools are used. Natural conditions are not distorted, and so the observations translate directly into statements about nature. In this tradition I put the anatomists and physiologists from Galen to Harvey. Although a certain amount of experimentation was used, by both Galen and Harvey, it was not at the center of this method. Had it been, one might have expected Vesalius, for example, to try to force fluid through the interventricular septum by means of a force pump in order to answer the question about the pores.[4] Although William Gilbert performed many simple experiments, they were little more than observations; nowhere did he transcend the discriminating power of the human senses or do violence to nature.[5] Francis Bacon was, of course, the great champion of meticulous and repeated observation in the seventeenth century.

In 1609 observational science suddenly attained an entirely new dimension, however. Simple observations with a new device, the telescope, unveiled hitherto unsuspected phenomena in the heavens. Following the publication of Galileo's *Sidereus nuncius* in 1610, there was a brief but very important controversy. Lenses and mirrors had been the playthings of dabblers in magic who stressed the miraculous effects achieved by means of them.[6] How could a combination of such distorting devices reveal nature as it really is? Fortunately, there was a solution to this problem which satisfied most conservatives. Telescopes showed things on earth the way they demonstrably were. At a feast given in his honor by Prince Federico Cesi in April 1611, Galileo let everyone inspect the inscription of Pope Sixtus V above the entrance of the Lateran Building, several miles away, through his telescope while it was still light. Then after dark he showed them the satellites of Jupiter.[7] Shortly afterward the mathematicians of the Collegio Romano certified to Cardinal Bellarmine, their spiritual director and the leading theologian in Rome, as well as a member of every Congregation of the Holy Office, that the things revealed by the telescope really did exist in the heavens.[8]

Obviously, most scientists will be skeptical and reserve judgment upon hearing a report about a new instrument that reveals entirely new phenomena in a long-familiar area of nature; and we would expect no less from reasonable natural philosophers in 1610. Upon seeing the newly discovered phenomena for themselves, however, and performing a few simple tests with the telescope, most if not all were convinced. This is not the whole story, though, because these new phenomena, such as Jupiter's satellites, required a mysterious amplification of the human visual power to be perceived. For the first time a set of phenomena had been added to science which required the mediation of an instrument to be detectable. Thinkers had to get used to this notion and mull over the implications. Moreover, it was not long before the telescope was joined by the microscope;[9] and in this case verification of the reality of the newly revealed phenomena was even more difficult. Had the way not been cleared by the telescope, surely the microscope would not have been accepted as quickly as it was. Henceforth the study of nature included the study of dimensions only accessible through optical aids. No two devices are more symbolic of the new science; they presented scientists with entirely new phenomena to be observed or manipulated.

Instruments must be seen in the context of the science done with them, and this context usually grows around the instrument. In the case of the telescope, Galileo's celestial discoveries initially had the status of revelations. One used a telescope to verify the discoveries and show them to friends. Few areas of further research readily suggested themselves. The study of sunspots[10] and the determination of the periods of Jupiter's satellites were the first.[11] To them was added the mapping of the moon in the 1630s;[12] but the practice of astronomy changed surprisingly little during the early decades of the telescope. The role of the astronomer remained almost exclusively the measuring of positions. Many astronomers did not even have good telescopes.[13] Only after 1640 can we say that all astronomers routinely observed the heavens with research telescopes, and not until a decade later do we notice a belief that better telescopes will reveal new phenomena in the heavens.[14] Only at that point had telescopic astronomy come into its own.

The development of the microscope presents an analogous picture. Early observations were curiosities. It stands to reason that an instrument that magnifies small things sufficiently can show a fly as big as an elephant,[15] but the step from curiosity to scientific investigation was made only when men began to study the anatomy of the fly for its own sake or for the sake of comparative anatomy. Microscopical studies began very slowly. Francesco Stelluti's study of the bee appeared in 1625,[16] Giovanni Battista Odierna's *Eye of the Fly* appeared in 1644,[17] and Pierre Borel's "century" of microscopical observations appeared in 1656.[18] Robert Hooke's *Micrographia* of 1665 finally signaled the beginning of a stream of monographic literature on this subject.

Only in the second half of the seventeenth century, then, did the microscope acquire a context of its own. The growth of a context around the instrument was, of course, not limited to the telescope and the microscope, the new instruments in observational science; it applied to all the new instruments to a greater or lesser degree. In short, new instruments tend to create new branches of science.

In the slow growth of the new science done with these instruments, internal technical factors cannot be ignored. The "Galilean" or "Dutch" telescope and the early form of microscope were not ideal research instruments.[19] Not until optical systems consisting of convex lenses only became common after 1640 did the telescope and microscope become convenient research instruments.[20] (This convenience must of course be measured by seventeenth-century standards.) Yet, clearly the drive was toward more power, and the very long telescopes of the second half of the seventeenth century as well as the awkward single-lens microscopes of Leeuwenhoek are proof that convenience was sacrificed for higher magnifications. The limiting factor, right from the beginning, was the level of technology. It was not until about 1650, for instance, that the major impediment, the difficulty of grinding long-focus objectives for telescopes, began to be overcome.[21] Glass quality remained a major stumbling block.[22] It is fair to say that right up to the twentieth century the slowly advancing technological front has been the limiting factor to progress in telescope and microscope construction and exploitation.

PRECISION MEASUREMENT

From the sources we now have in translation, it appears that the very accurate parameters of the motions of the planets, the sun, and the moon current among the Mesopotamians after about 700 B.C. were the result not of accurate measurements but rather of very mediocre measurements continued over a long period of time.[23] We must go to late antiquity, the period from Hipparchus to Ptolemy (ca. 150 B.C.–ca. A.D. 150), to find the first attempts at accurate and convenient measurement to solve specific astronomical problems.[24] Even in the case of Ptolemy, however, it is obvious that he relied heavily on Mesopotamian sources.[25]

In Europe, serious consideration of the entire Ptolemaic scheme with a view toward greater accuracy did not begin until the late fifteenth century, with the work of Peurbach, Regiomontanus, and Walther.[26] The work of Copernicus must be seen as an attempt to restore astronomical theory to a more appropriate basis, one without an equant.[27] Not until Tycho Brahe do we see a coordinated attack on the problem of observational accuracy; but Tycho's aim, too, was to *restore* astronomy.[28] It is difficult to view his instruments as major technological innovations, although, to be sure, his scale divisions, calibrations, and methods of observation represented important breakthroughs in technique.[29]

In measuring the positions of heavenly bodies, two important motivations must be separated: convenience and accuracy. The dioptra of Hipparchus and Hero was an attempt to achieve greater accuracy;[30] the astrolabe sacrificed accuracy for convenience.[31] The medieval European instruments all aimed toward convenience[32] (a sure sign of the derivative nature of medieval astronomy). In astrology, too, the accent was never on observational accuracy. Knowing roughly the positions of the planets and the luminaries at the moment of someone's birth or a few months in the future was enough for a horoscope. If there was a problem in astrology it was one of longer-range predictions, and this problem involved theory—the basis of tables—rather than observation.[33]

Tycho Brahe sacrificed convenience to accuracy, with the result that his observations were unprecedented in their accuracy.[34] In the sixteenth century there was a dawning realization that the surviving observations from Ptolemy onward were of greatly varying accuracy and that in fact the observations made in antiquity were not nearly as accurate as Ptolemy's great authority would predispose one to believe. Tycho rejected all observations before the sixteenth century and hoped to reform all of astronomy with his very accurate observations during his lifetime.[35]

Beyond that, however, the notion that today's accuracy is tomorrow's error, that every generation must bring the observational data base in line with new standards of accuracy, is utterly lacking in Tycho and his immediate successors. His observations allowed Kepler to reform astronomical theory, and only a few problems, it seemed, were left to be mopped up after 1630—problems such as lunar theory and the relationship between refraction and solar parallax, which affected the value of the obliquity of the ecliptic.[36] But Tycho's successors in observational astronomy did not think of superceding his observations, only of completing them, and therefore there were few serious efforts to build instruments that would rival those of Tycho in accuracy. Only the sumptuously equipped rooftop observatory of Johannes Hevelius in Gdansk could vie with Uraniburg;[37] and Hevelius explicitly rejected the suggestion of improving on Tycho's accuracy by means of instruments equipped with telescopic sights which were introduced in the 1660s.[38] This new practice was instituted by the younger generation of John Flamsteed, Jean Picard, and Ole Roemer.[39] It was this revolution in astronomical accuracy, beginning around 1660, that finally made astronomers aware of the inadequacy of all previous observations, including those of Tycho Brahe.[40]

The story with clocks is similar. Early water clocks were sophisticated but not accurate.[41] Weight-driven clocks and then spring-driven clocks represented a great advance in convenience but not in accuracy.[42] Astronomical measurement remained confined to direct measurements of angular separations. Although Tycho Brahe had several clocks in his house and observatory, they played no role in his measurements because they were too irregular even over short time intervals.[43] Not until the pendulum was fitted to the weight-driven clock as a regulator, after 1656, did clocks become sufficiently accurate for astronomical purposes.[44] Pendulum clocks quickly became standard equipment in all serious observatories. Differences in right ascensions could now be measured with clocks, and reasonable separations in right ascensions could be measured accurately to considerably less than one second of time, which corresponds to fifteen seconds of arc. Transit instruments now became the centerpieces of all major observatories.[45]

When Tycho began accumulating his celebrated instruments, although he was making new demands on the skills of the artisans, their construction lay well within the general technological capability of the time. The basic components were there, but Tycho had craftsmen put them together in a new way. He made the initial shift from wood to metal because of the greater inherent precision of metal components. Aside from that, he obtained unprecedented accuracy by making his instruments larger and paying much more attention to detail than his predecessors had ever done.[46] Likewise, the quick spread of pendulum clocks and the rapid improvement of them between 1660 and 1700 indicates that in this, too, general technology was not a limiting factor up to the eighteenth century.[47]

By 1700 the measurement of time and space intervals in astronomy had taken on the basic form it was to retain for several centuries. Accuracies transcending those of the human senses by several orders of magnitude had become common in branches of science in which such measurements were of consequence. The notion that man could seriously investigate the heavens with the unaided senses had become obsolete. Having outlined the important changes in simple observation and precision measurement, I shall now turn to the additional impact on those practices of the demands of two new and interrelated methods of science, experimental and mathematical.

MATHEMATICAL AND EXPERIMENTAL SCIENCE

It would be a mistake to say that mathematics was first applied to nature in the late sixteenth century. One only has to think of the Pythagoreans, Archimedes, and Ptolemy to realize the importance of the antecedents. But Archimedes addressed himself to problems of statics only;[48] Ptolemy's geometrical constructions were considered mere devices to "save the phenomena," not descriptions of reality;[49] and the fourteenth-century mathematical treatments of changing qualities were formulated *secundum imaginationem*, not *secundum cursum naturae*.[50] It was only with the disintegration of the Aristotelian cosmos, the identification of space with the mathematical space of Euclid, and the consequent conviction that nature is mathematical—changes that began in the sixteenth century—that one can begin to speak of mathematical science in the way we understand it today.[51] The pivotal generation was that of Kepler, who introduced physics into the heavens, and Galileo, who argued that mathematics describes real processes on earth.

Likewise, experiments had been performed in antiquity and the Middle Ages. Hero of Alexandria showed that open vessels submerged in water will not fill with water unless air is allowed to escape;[52] Galen showed by means of ligatures around the ureters and urethra of a living animal that urine flows from the kidneys to the bladder through the ureters;[53]

and Theodoric of Freiburg investigated the nature of the rainbow by permitting the sun's rays to pass through glass vessels filled with water.[54] These are isolated examples, however. To ascribe experimental science to the period before 1600 on the basis of isolated experiments is analogous to calling the medieval economy a market economy because there were markets in the various settlements. Only when it was recognized that experimentation is an essential part of investigating nature, around 1600, can we begin to recognize a continuous tradition of experimental science.

The legitimacy and usefulness of scientific instruments, taken for granted today, derive finally from their integrated status in mathematical and experimental science. The relationship between instrument and science did not spring forth fully formed around 1600; it grew in the course of the seventeenth century. Thus, Santorre Santorio (1561–1636), an iatro-mathematician with a passion for measurement, constructed a *pulsilogium* (a small, hand-held pendulum) to measure his patients' pulses, devised a thermoscope with which to measure the degree of heat of the human body, and even performed the first quantitative metabolic experiments using himself as subject.[55] His efforts, however, remained curiosities, unintegrated in the scientific theory of his time, and although thermometers became more and more elaborate in the course of the century, their usefulness was severely restricted by a lack of standardization and an inadequate theory of heat.[56]

If measurement was entirely new to the science of medicine, it had played an important role in astronomy since antiquity. Yet here, too, theory had to legitimize the measurements of unprecedented accuracy made by Tycho Brahe. Rather than cite the well-known case of Kepler's elliptical orbits, let me give a different illustration. Tycho had recognized that the sun is apparently displaced upward as it approaches the horizon. He ascribed this phenomenon to the refraction of light by the vapors close to the earth, and he instituted a program of measuring, with his very accurate instruments, the elevations above the horizon of the sun, moon, and fixed stars. Comparing these measurements with predictions, he concluded that refraction was insensible above forty-five degrees in the cases of the sun and moon, and above twenty degrees for the fixed stars and planets. His measurements translated directly into three different refraction tables: one for the sun, one for the moon, and one for the fixed stars and planets.[57] Clearly, accurate measurements by themselves do not help science very much.

Kepler recognized the inherent contradiction in having three different refraction tables and devoted an entire treatise, his *Astronomia pars optica* (1604), to the problem of refraction. He started from first principles and argued that refraction was due to the different densities of ether and atmosphere, that it continued right up to the zenith, and that it had to be the same for all heavenly bodies. In his own tortuous way, he searched for a mathematical relationship based on these principles, one that would be commensurate with known measurements, including Tycho's. The result was his refraction law or formula, which although erroneous in retrospect, gave good results for altitudes above ten degrees. Kepler was then able to construct a single refraction table for all heavenly bodies, more accurate by our standards than Tycho's.[58]

Whereas Tycho had been satisfied with ad hoc corrections for the elevations of heavenly bodies, Kepler tried to put the entire science of refraction on a mathematical footing, using Tycho's accurate measurements. His approach served to focus attention on one of the main problems in observational astronomy in the seventeenth century: how could one explain the apparently different refraction corrections arrived at by Tycho? As it turned out, the solution to this problem lay in the relationship between refraction corrections and corrections for solar parallax. Here, theory pointed the way to even more accurate measurements. If mathematical science must be based on a foundation of accurate measurements, it in turn makes demands for further accurate measurements. Once this mutually reinforcing relationship had been established, the demand for accurate instruments escalated.

Whereas Kepler tried to establish mathematical science on the basis of previously existing measurements, Galileo's case is more complicated. It is clear that Galileo performed both qualitative and quantitative experiments on moving bodies and at the same time tried to derive a new mathematical science of motion from first principles. The relationship between these two methods presents us with a historiographical problem, however. Paradoxically, in the transition from Aristotelian to classical science, science became at once more practical and more abstract. In studying Galileo, historians (with few exceptions) have stressed the abstraction and subordinated the practicality. Alexandre Koyré was the guiding spirit of this movement. In his important studies on Galileo he argued that dropping weights from the leaning tower of Pisa could prove very little, that inclined-plane experiments could not possibly produce the law of falling bodies, and that only in the abstract Euclidean space of his own mind could Galileo roll (perhaps I should say slide) perfectly hard and spherical balls down perfectly smooth inclined planes.[59] According to Koyré, Galileo was a Platonist who identified the real with the ideal, and it was this leap into the abstract world which allowed him to find the law of falling bodies which no amount of experimentation could have revealed:

Good physics is made *a priori*. Theory precedes fact. Experience is useless because before any experience we are already in possession of the knowledge we are seeking for. Fundamental laws of motion . . . , laws that determine the spatio-temporal behaviour of material bodies, are laws of a mathematical nature. Of the same nature as those which govern relations and laws of figures and numbers. We find and discover them not in Nature, but in ourselves, in our mind, in our memory, as Plato long ago has taught us.[60]

Koyré was followed in his Platonism by a generation of scholars. According to I. B. Cohen, Galileo found the laws of motion with, at best, the help of pen and paper, and his experiments were "only . . . a kind of rough check to see whether the principles that he had derived by the method of abstraction and mathematics actually applied in the world of nature."[61] N. R. Hanson even went so far as to say: "Centuries of scholarship to the contrary notwithstanding, Galileo was not a great experimental scientist. He was no experimental scientist at all, not as we would know one."[62] Needless to say, Koyré's thesis has had a profound influence on the historiography of the Scientific Revolution.[63]

In the past two decades, Stillman Drake has patiently dated and reconstructed the meanings of a number of unpublished Galileo manuscript pages and scraps.[64] Drake has lately supplied convincing evidence that Galileo performed experiments with inclined planes before he had found the law of falling bodies, that these experiments were sufficiently accurate to give useful results for distances traveled during successive equal time intervals, and that, in fact, Galileo found the time-squared law as well as the parabolic trajectory by means of these experiments![65] From the very beginning, then, experiment played an important role in the *discovery* of the laws of motion. Historians have been misled by Galileo's finished demonstrations in which experiments are introduced only to confirm previously established mathematical relationships.[66] Galileo was, in fact, much more practical than has been supposed. Abstraction and practicality went hand in hand, and mathematical and experimental science grew up together. Scientific instruments bridged the gap between them.

I am not arguing that the laws of nature can be established by an unprejudiced Baconian approach. Clearly, Galileo had to resolve some very difficult conceptual problems before he could arrive at the correct relationship. He found the solution to the problem in his own mind. But I do argue that in his search he obtained crucial information from nature by questioning her by means of experiments—quantitative experiments. The science of motion, beginning with Galileo, was a mathematical and experimental science. Experiments played an important role in the *discovery* of relationships, but their role was minimized in the *demonstration*. Mathematical demonstrations were certain but the role of experiment remained problematical in the physical sciences until somewhat later. To put it a different way, the relationship between mathematics and nature was accepted as a matter of course by Galileo and his successors, while the relationship between experiment and natural law remained more subject to debate.

If the book of nature is written in the language of mathematics, then numbers are needed to change proportions into absolute numerical equations. In the heavens these numbers had been obtained by observations made since time immemorial, but on earth the situation was different. In the science of motion, distance could be measured quite accurately, but small intervals of time were a great problem. Using inclined planes to reduce the velocity of a falling body, Galileo found the relationship between average velocities over successive time intervals, a relationship that eventually gave him the proportionality between distance and time-squared. This proportion applied to falling bodies, but how fast do bodies actually fall? Inclined planes and water clocks could not help answer this question: they were too inaccurate. It appears that Galileo never made this measurement,[67] but Marin Mersenne and Giovanni Battista Riccioli did. Their methods of measuring time, using the half swing of a pendulum and rhythmic counting, were extremely difficult and not very accurate. Yet, despite these obstacles, Mersenne was able to conclude that a body falls about twelve (Parisian) feet in the first second, and Riccioli found it to be about fifteen (Roman) feet.[68] The mathematical investigations of the truly isochronous pendulum led Christiaan Huygens to the formula relating pendulum length and period to the acceleration due to gravity, and thus he was able to obtain essentially the modern figure for the acceleration due to gravity.[69] The pendulum and the pendulum clock were instruments based from the beginning on scientific principles. In fact, since the measurement of the acceleration due to gravity with a pendulum is indirect, the instrument would in this case be meaningless without the underlying physical and mathematical theory.

The barometer was, likewise, an instrument that grew directly out of scientific principles and was legitimated by them. The observation that suction pumps can only raise water to a height of about thirty feet and that siphons also fail to operate at greater heights led to investigations of this phenomenon in Rome in the 1630s.[70] From the very beginning, various theoretical explanations were suggested. Giovanni Battista Baliani argued that the weight of the atmosphere was the causal mechanism, whereas Galileo invoked the notion of a "rope of water" breaking under its own weight.[71] The experiments of Torricelli and Pascal are well known. Torricelli is credited with the invention of the barometer because he set out to measure the weight of the atmosphere.[72] The very nature of the barometer made it easy to standardize (although I do not wish to underestimate the problems caused by the variety of local weights and measures).[73]

The Torricellian experiments could not settle the argument between the "vacuists" and "plenists."[74] Independent detection of subtle fluids or an ether remained a problem until Einstein banned it from physics; but the mercury column experiments led directly to the invention of the air pump, and the air pump provided experimenters with a convenient artificial environment in which to experiment. It was the artificiality of this environment, however, that raised an age-old problem, the opposition of art and nature. According to the traditional view, art could at best produce only poor imitations of nature, and the artificial heat of the alchemist's furnace was therefore different from, and inferior to, the natural heat of the sun.[75] Could a totally artificial environment be useful in the study of nature? Bacon and his followers emphatically answered this question in the affirmative.[76] Thus, Henry Power wrote in his *Experimental Philosophy*:

For Art, being the Imitation of Nature (or, Nature at Second-Hand) it is but a sensible expression of Effects, dependent on the same (though more remote Causes;) and therefore the works of the one, must prove the most reasonable discoveries of the other. And to speak yet more close to the point, I think it is no Rhetorication to say, That all things are Artificial; for Nature it self is nothing else but the Art of God.[77]

Robert Boyle's experiments with the air pump were the first to be performed in this new, totally artificial and controlled environment. And whereas Pascal, in his *Treatise on the Equilibrium of Liquids* (written in 1654 but not published until 1663), had followed the mathematical approach of exposition, in which propositions were illustrated or demonstrated by sketchily described real experiments and thought experiments,[78] Boyle allowed real experiments to speak for themselves. His detailed descriptions of apparatus and procedure allowed others to repeat his experiments and verify the results, though the explanation of these results might remain a matter of debate. Thus, in his *New Experiments Physico-Mechanical Touching the Spring of the Air*, of 1660, Boyle demonstrated the "spring" of air by a number of experiments but made it clear that the explanation of this spring was not the subject of the treatise. His business was not "to assign the adequate cause of the spring of the air, but only to manifest, that the air hath a spring, and to relate some of its effects."[79] He saved causes for his more philosophical treatises.

Newton went a step farther. In his paper of 1672 he based an entirely new theory of light and colors on a few carefully chosen and well-executed experiments. If Newton felt that his explanation of colors was a "rigid consequence" of his experiments,[80] however, others disagreed.[81] And although in the long run Newton overcame his critics, the relationship between experiment and theory remained problematical.[82] Granted that a theory can be supported or refuted by experiments, it is by no means clear how a theory can follow from or be produced by experiments.

As we have seen, observational instruments, measuring instruments, and experimental instruments were shaped by the demands of the new mathematical and experimental methods of science. It is time now to attempt some generalizations.

THE SCHOLAR AND THE CRAFTSMAN

In a scientific environment in which the types and numbers of scientific instruments were proliferating rapidly, a cast of suitable artisans was necessary. The relationship between scientists and the growing community of increasingly sophisticated craftsmen has been the subject of some studies. On the whole, the conclusion has been that the initiative in their contact came from the scientists, and I agree with this conclusion.[83] Like some other areas, science acquired a technology in the seventeenth century, and as science moved from the library to the laboratory, it became increasingly dependent on a cast of specialized craftsmen, the instrument makers.

These craftsmen were not readily available in 1600. Although there were skilled clock makers who worked for highly placed patrons in the sixteenth century (Yost Bürgi, for instance[84]), they were very much the exception. The average craftsman[85] worked in well-defined and limited areas, and he was governed more often than not by guild regulations that discouraged deviation from established practices and product lines. On the whole, the craft structure in 1600 was much too inflexible to deal with the changing demands of the scientific community.[86] The spyglass, for instance, came from the craft tradition.[87] To produce it spectacle makers put two lenses, existing products, in a tube. But the improvement of this gadget into the research telescope was quite a different matter. For this purpose very high quality glass and lenses with strengths well outside the range used for ordinary spectacles were required. The glassmaking craft was very slow to respond, and the spectacle-making craft never adjusted to the new demands.[88] Slowly, a specialty craft of telescope and microscope makers developed in the course of the seventeenth century,[89] but until the very end of the century scientists competed successfully with telescope makers. And it is interesting to note that in several instances scientists were able to do what artisans could not. Thus, Galileo was able to improve the telescope well before craftsmen could,[90] and two excellent telescope makers, Richard Reeves and Christopher Cock, were unable to make a mirror telescope for the mathematician James Gregory in 1663,[91] whereas Newton working by himself succeeded in his efforts a few years later.[92] I suggest that in the seventeenth century, under the spurs of scientists, a body of craftsmen grew up who were not bound by tradition and who learned to push the frontiers of technology and technique continuously ahead.

Not every scientific instrument was always held back by the limits of technology. There is, for instance, little evidence that craftsmen had much trouble making the measuring arcs with telescopic sights for Jean Picard and his colleagues at the Royal Observatory in Paris in the 1660s.[93] Nor was there a particular technological barrier to the execution and improvement of the pendulum clock immediately after 1660.[94] It was only at the beginning of the eighteenth century that instrument makers across the board were working at the very frontiers of precision technology.[95]

The size of the community of instruments makers, and hence its quality, was limited by demand and grew only slowly. For the time being, rich patrons continued to play an important role in the support of highly skilled craftsmen. High-quality instruments were always expensive, and even when a scientist needed an instrument there was no guarantee that he could pay for it.[96] No doubt, this is one of the reasons scientists such as Galileo, Huygens, and Hooke made their own instruments. In addition, the ability to make one's own instruments, an excellent research telescope, for instance, had become a mark of prowess in an increasingly competitive practical scientific community.[97] This tradition has had a long life, and even today experimental scientists and their technicians routinely make new pilot instruments for specific jobs.

ANCIENTS AND MODERNS

Though of unprecedented accuracy, Tycho's instruments and observations had as their purpose the restoration of astronomy, and it is wrong to read into them the notion of continued increases in accuracy over time. Likewise, as startling as Galileo's telescopic discoveries were, it is a mistake to assume that the notion of ever more discoveries with ever more powerful telescopes was obvious in 1610. In both cases the only obvious road for the future was the solution of remaining problems. The status of Tycho's observations and Galileo's discoveries was akin to that of divine revelation.

When he was apprised of the new and more powerful telescopes of Francesco Fontana in 1638, Galileo wrote to his disciple Benedetto Castelli that these instruments merely magnified more but revealed nothing new. He had discovered it all long ago.[98] Even after Christiaan Huygens had discovered a satellite of Saturn in 1655,[99] the first major discovery since sunspots in 1611, Christopher Wren could write, in 1658:

The incomparable Galileo, who was the first to direct a telescope to the sky—although the telescope had then only recently been invented and was not yet in all respects perfected—so overcame yielding nature, that all celestial mysteries were at once disclosed to him. . . . His successors are envious because they believe that there can scarcely be any new worlds left, about which they can boast, and believe that only to succeeding Lyncei is it granted to add to the discoveries of Galileo.[100]

The Accademia de' Lincei of Prince Federigo Cesi was long dead,[101] and the reference to Lincei, or lynx-eyed men, was obviously meant to exclude mere mortals.

The telescope and microscope were, of course, used by the "moderns" as evidence of the superiority of their age over antiquity. Many examples of this use of these instruments can be cited.[102] But this was not the same as saying that tomorrow's improved instruments would bring yet more discoveries. We

find this notion expressed with unimpeachable clarity for the
first time in Hooke's *Micrographia* of 1665:

'Tis not unlikely, but that there may be yet invented several other helps for
the eye, as much exceeding those already found, as those do the bare eye,
such as by which we may perhaps be able to discover *living Creatures* in the
Moon, or other Planets, the *figures* of the compounding Particles of matter,
and the particular *Schematisms* and *Textures* of Bodies.[103]

In the second half of the seventeenth century, then, we see
something approaching the idea of progress applied to scientific instruments, a notion that today we take for granted. The
notion of restoration and reform was being replaced by that of
continuing growth.[104]

In a slightly different vein, the controversy between Hevelius and the two Englishmen Hooke and Flamsteed represents
the watershed in precision measurements. Hevelius defended
open sights on measuring arcs because he did not think that
the accuracy of Tycho's measurements could be surpassed.[105]
Hooke and Flamsteed defended the use of the recently introduced telescopic sights. Flamsteed discovered errors of several
minutes of arc in the measurements of Tycho and Hevelius,
and in his opinion only measuring arcs with telescopic sights
could help rid astronomy of such errors.[106] He (at Greenwich
starting in 1675) and his colleagues at the Paris Observatory
began to recast observational astronomy completely according
to the new standards of accuracy. Although we still find vener-
ation of his great predecessors in observational astronomy in
Flamsteed's utterances, it is clear that *authority* had been trans-
ferred from past generations to the present generation, which
possessed entirely new and much more accurate instruments.

CONCLUSION: THE COMPLETE NOTION OF
A SCIENTIFIC INSTRUMENT

By the end of the seventeenth century, science without
instruments had become inconceivable. In the biological sciences the microscope had revealed phenomena completely
unknown to the ancients—for instance, animalculae and capil-
laries in the lung of a frog. And even the investigations by the
ancients of such phenomena as the formation of the chick in
the egg or the workings of a bee colony had been made

completely obsolete through microscopical investigations.[107] In astronomy, some of the most important investigations, such as those of the periods of the satellites of Jupiter and Saturn, were impossible without the telescope, and the implications for universal gravitation of the proof that these bodies obeyed Kepler's third law are readily apparent.[108] It is fair to say that, whereas in 1600 scientific instruments had only limited roles in science (e.g., in astronomy), by 1700 instrumentation had become an accepted dimension of all of science. It was now a commonplace that instruments had an essential place in the study of nature.

In 1600 the determinative nature of instrumental readings had barely been established by the measurements of Tycho Brahe. One only has to recall the widely varying parallax measurements of the new star of 1572 and the comet of 1577 to realize that it took the revolution instituted by Tycho to drive home the point that perhaps a theory could be shown wrong by a measurement.[109] Clearly we are not justified in saying that in 1550 the readings of instruments were beyond dispute. The telescope was the first instrument to force a widening of trust in instrumental reliability; and although the reality of the celestial phenomena revealed by Galileo was quickly established, before 1650 the quality of telescopes was such that distortion remained a clearly evident problem to all involved. Only the improvement of the instrument and the parallel development of the science of optics, which gave theoretical explanations of distortion and color fringes, could change this. In his Gresham lectures of 1681, John Flamsteed still felt it necessary to establish by detailed analysis of optical systems that lenses do not "impose upon our senses."[110]

The Torricellian tube, but even more the air pump, first established a completely artificial environment—an environment impossible according to Aristotle. Whether one was a vacuist or a plenist, the proof that sound needs air to be transmitted and that light does not[111]—and the proof of Galileo's contention that, in the absence of the retarding influence of air, an air-filled bladder falls as fast as a lead ball of the same size[112]—these were crucial experimental contributions that helped erase the opposition of art to nature and to establish the validity of a controlled experimental environment. Needless to say, proper experimental technique and reproducibility of results were crucial in this respect, and the work of Robert Boyle cannot be praised enough in this context.

The relationship between instrument and theory was by no means obvious at the beginning of the seventeenth century. When Galileo published *Sidereus nuncius* in 1610, there was no adequate theory of lenses. Beginning with Kepler's *Dioptrice* in 1611, determined efforts were made to supply this new instrument with a theoretical foundation. By the end of the century, through the work of Descartes, Huygens, Newton, and Molyneux, the telescope had been put on a firm theoretical foundation. In the case of the pendulum, its isochronous nature was an erroneous article of faith with Galileo. Not until Huygens was the instrument supplied with an adequate theoretical basis. Huygens's treatment was so elegant that it established the pendulum clock as the theory-based instrument par excellence of the seventeenth century. No instrument inspired greater faith in the validity of instruments.

The pendulum was also a clear example of the circularity of science, however. The theory on which it was based was the new mathematical physics of Galileo; to an Aristotelian it could have little meaning. And we should, of course, add the air pump to this list. In this sense, the relationship between instrument, experiment, and theory remained problematical.

We come finally to the problem of *generations* of instruments. As we have seen, the idea of improving on Tycho's accuracy and on the discoveries of Galileo did not suggest itself as readily as we might suppose. Not until Huygens's discovery of a satellite of Saturn in 1655 was the telescope race on in earnest, and not until after 1660 were realistic hopes entertained of improving on Tycho's measurements across the board. The competition for celestial discoveries with increasingly powerful telescopes, instruments that grew to absurd lengths,[113] is the very first instance of the instrument races that have become common today. Individual and national prestige were on the line. The French, with the help of a naturalized Italian astronomer, Jean Dominique Cassini, and a Roman telescope maker, Giuseppe Campani, easily won this race because of their superior endowment of science.[114] Good instruments were never cheap.

A more important revolution in precision measurement of

planetary and stellar positions came with the application of telescopic sights to measuring arcs. It is difficult to overestimate the importance of this revolution. The French completely rewrote the geography book of their country before 1700. Errors that are almost inconceivable by modern ideas were uncovered in the separations between major cities. (Louis XIV complained that astronomers were costing him more territory than his foreign enemies.)[115] Tycho was shown to have erred by two minutes of arc in the obliquity of the ecliptic, and the size of the solar system was expanded by a factor of about twenty.[116] In England, Flamsteed and his successors, especially Bradley, continuously improved their instruments, and Bradley's discovery of the aberration of light and of nutation are eloquent testimonies to the accuracy of his instruments.[117] These variations were so small that they were completely swamped by the error margins of instruments with open sights. Now, each generation consciously set out to improve instruments in close cooperation with instrument makers, and thus there began to develop a generational notion of accuracy and of instrument quality and power.

It was only when the tacit assumptions we now have relative to scientific instruments were generally accepted—that is, when scientific instruments had obtained an unquestioned role in science, when their results and readings were no longer subject to questions beyond those of repeatability and standards of accuracy, when they had been firmly based on agreed principles, and when continuous improvement had become the quest of the community of scientists and instrument makers—that we can say the modern notion of a scientific instrument had been born.

NOTES

1. A *New English Dictionary on Historical Principles*, 10 vols. (Oxford: Clarendon Press, 1888–1928), 5:356–357:

A material thing designed or used for the accomplishment of some mechanical or other physical effect; a mechanical contrivance (usually one that is portable, of simple construction, and wielded or operated by the hand); a tool, implement, weapon.

Now usually distinguished from a *tool*, as being used for more delicate work or for artistic or scientific purposes: a workman or artizan has his *tools*, a draughtsman, surgeon, dentist, astronomical observer, his *instruments*. Distinguished from a *machine*, as being simpler, having less mechanism, and doing less work of itself; but the terms overlap.

In this sense the word was already used by Chaucer in 1391 to describe the astrolabe. See R. T. Gunther, "Chaucer and Messahalla on the Astrolabe," in Gunther, *Early Science in Oxford*, 14 vols. (Oxford: Printed for the subscribers at the University Press, 1923–1945), 5:1.

2. I am not suggesting that measurements have meaning outside the framework of a theory. For a discussion of the relationship between theory and measurement, see Thomas S. Kuhn, "The Function of Measurement in Modern Physical Science," *Isis* 52 (1961): 161–190; reprinted in Kuhn, *The Essential Tension: Selected Studies in Scientific Tradition and Change* (Chicago: University of Chicago Press, 1977), pp. 178–224. It is the validity of results or measurements within the framework of a theory which I am concerned with here.

3. The balance had, of course, always been part of the equipment of alchemists and chemists—Van Helmont, Boyle, and Black performed quantitative analyses with it—but it was Lavoisier who first stated the law of conservation of matter as the fundamental principle of chemistry:

> We may lay it down as an incontestible axiom, that, in all the operations of art and nature, nothing is created; an equal quantity of matter exists both before and after the experiment; the quality and quantity of the elements remain precisely the same; and nothing takes place beyond changes and modifications in the combination of these elements. Upon this principle the whole art of performing chemical experiments depends: We must always suppose an exact equality between the elements of the body examined and those of the products of its analysis.

(*Traité élémentaire de Chimie* [1789], trans. Robert Kerr, *Elements of chemistry* . . . [Edinburgh, 1790; reprint ed., New York: Dover, 1965], pp. 130–131).

4. For a convenient and accurate translation of Vesalius's statements concerning the pores of the septum in both the 1543 and 1555 editions of *De humani corporis fabrica* (Book VI, chap. 11), see Gweneth Whitteridge, *William Harvey and the Circulation of the Blood* (London: MacDonald; New York: American Elsevier, 1971), p. 46. Vesalius appeals only to the unaided senses.

5. For example, in Book II, chap. 2 of *De magnete*, Gilbert reports the results of his electrical investigations, carried out with the help of a nonmagnetic versorium (a piece of metal on a pin). The thirty-three discoveries marked by asterisks are all based on very simple experiments that are not much more than observations. See *On the Magnet*, trans. Sylvanus P. Thompson (London, 1900; reprint ed., New York: Basic Books, 1958), pp. 46–60; Duane H. D. Roller, *The "De magnete" of William Gilbert* (Amsterdam: Menno Hertzberger, 1959), pp. 112–114.

6. For pre-1600 claims about the miraculous effects of lenses and mirrors, see Albert Van Helden, *The Invention of the Telescope, Transactions of the American Philosophical Society* 67, part 4 (1977): 12–16, 28–35. The problem of vision through optical systems and the difficulties of locating the effigies thus produced in the mind are discussed by Vasco Ronchi in *Optics: the Science of Vision*, trans. Edward Rosen (New York: New York University Press, 1957), pp. 124–204. See also Paul K. Feyerabend, "Problems of Empiricism," Part II in *The Nature and Function of Scientific Theories*, ed. Robert G. Colodny (Pittsburgh: University of Pittsburgh Press, 1970), pp. 275–353, esp. pp. 281–289; and Peter K. Machamer, "Feyerabend and Galileo: The Interaction of Theories, and the Reinterpretation of Experience," *Studies in History and Philosophy of Science* 4 (1973–1974): 1–46. See also Terrie F. Bloom, "Borrowed Perceptions: Harriot's Maps of the Moon," *Journal for the History of Astronomy* 9 (1978): 117–122.

7. Edward Rosen, *The Naming of the Telescope* (New York: Henry Schuman, 1947), pp. 53–54.

8. *Le opere di Galileo Galilei*, Edizione Nazionale, 20 vols. (Florence, 1890–1909, reprinted 1929–1939, 1964–1966), 12:87–88, 92–93.

9. The claim that Sacharias Janssen and his father Hans had invented the compound microscope in Middelburg in 1590 was first made by J. H. Van Swinden in the 1820s. See Gerard Moll, "Geschiedkundig Onderzoek naar de Eerste Uitvinders der Verrekijkers uit de Aantekeningen van wijlen den Hoogleeraar Van Swinden zamengesteld," *Nieuwe verhandelingen der eerste klasse van het Koninklijk Nederlandsch Instituut van Wetenschappen, Letterkunde en Schoone Kunsten* 3 (1831): 171; idem, "On the first Invention of Telescopes," *Journal of the Royal Institution* 1 (1831): 319–332, 483–496. Van Swinden's student Pieter Harting was a very influential advocate of this point of view; see *Das Microskop* (Braunschweig, 1859), pp. 585–595. The claim is therefore still discussed in many books on the microscope, e.g., Alfred N. Disney, *Origin and Development of the Microscope* (London: Royal Microscopical Society, 1928), pp. 89–107; Reginald S. Clay and Thomas H. Court, *The History of the Microscope* (London: Charles Griffin, 1932), pp. 7–13; and S. Bradbury, *The Evolution of the Microscope* (Oxford: Pergamon Press, 1967), pp. 21–22. Gilberto Govi argued that Galileo was the first to make a compound microscope by adapting the Galilean telescope. See "Il Microscopio Composto Inventato da Galileo," *Atti del Reale Accademia delle Scienze Fisiche e Matematiche in Napoli* 2, series 2 (1888): 1–33. All the claims made for Sacharias Janssen, his father Hans, and his son Johannes Zachariassen, were examined in detail by Cornelis de Waard in *De uitvinding der verrekijkers* (The Hague, 1906). De Waard discounted the argument of Van Swinden and Harting and argued convincingly that the compound microscope grew out of the telescope and that Galileo already adapted his telescope for viewing small, nearby things in 1610 (pp. 293–304).

10. After the initial controversy between Christoph Scheiner and Galileo on the nature of sunspots, ending with the publication of Galileo's *Istoria e dimostrazioni intorno alle macchie solari* (Rome, 1613), Scheiner continued his study of the sun and presented his results in *Rosa ursina* (Bracciani, 1630). This exhaustive treatise was the standard work on sunspots for more than a

century, owing partly to the minimal sunspot activity during the so-called Maunder Minimum, 1645–1715.

11. It was quickly recognized that the eclipses of Jupiter's satellites might be used to determine longitude differences. As good telescopes became more common, this method was used successfully for longitude differences between fixed localities. For the more pressing problem of longitude at sea, however, tables of the eclipses were needed, and these required very accurate determinations of the periods of the satellites. The first fairly accurate tables were those of Giovanni Domenico Cassini in 1668. Not until the eighteenth century were there tables sufficiently accurate for determining longitude at sea, but then John Harrison's accurate chronometers presented sailors with a much more convenient method.

12. O. Van De Vyver, *Lunar Maps of the XVIIth Century, Vatican Observatory Publications*, vol. 1, no. 2 (1971).

13. Gassendi used a very poor telescope in 1619 and 1620 (*Opera omnia*, 6 vols. [Lyons, 1658], 4:82). In 1634 Claude Nicholas Fabri de Peiresc (Gassendi's patron) complained to Galileo about the poor quality of his telescopes and asked Galileo to send one of his (*Opere*, 16:28, 17:34–35). In 1637 the Dutch astronomer Martinus Hortensius likewise mentioned to Galileo the poor quality of Dutch telescopes (*Opere*, 17:19).

14. Observations of the moon, Mars, and Saturn, made with the new telescopes of Francesco Fontana, circulated in Italy as early as 1638. See Gino Arrighi, "Gli 'Occhiali' di Francesco Fontana in un Carteggio Inedito di Antonio Santini nella Collezione Galileiana della Biblioteca Nazionale di Firenze," *Physis* 6 (1964): 432–448. Jupiter's equatorial belts were first seen in Naples at about the same time. See Giovanni Battista Riccioli, *Almagestum novum*, 1 vol. in 2 parts (Bologna, 1651), 1:487. It was not until the late 1640s, however, that good pictorial representations of these phenomena in print became a practice among astronomers, e.g., Fontana, *Novae coelestium terrestriumque rerum observationes* (Naples, 1646); Johannes Hevelius, *Selenographia* (Gdansk, 1647); Riccioli, *Almagestum novum*. Not until about 1650, therefore, was there a communal sense of exactly what was new in telescopic observations and a realization that this branch of astronomy was beginning to improve on Galileo's discoveries.

15. Galileo related to Jean Tarde in 1614 that he saw flies as big as lambs (*Opere*, 19:590). Scheiner saw a fly as big as an elephant and a flea as big as a camel (*Rosa ursina*, p. 130^r).

16. *Apiarum* (Rome, 1625). Stelluti's *Persio tradotto in verso scioloto* (Rome, 1630) also contains magnified views of bees and a grain weevil. See Charles Singer, "The Earliest Figures of Microscopic Objects," *Endeavour* 12 (1953): 197–201.

17. *L'Occhio della mosca*, in *Opuscoli del Dottor Gio. Battista Hodierna* (Palermo, 1644).

18. *Observationum microscopicarum centuria* (The Hague, 1656). This work was issued jointly with Borel's *De vero telescopii inventore* (The Hague, 1655). See A. Van Helden, "A Note about Christiaan Huygens' *De Saturni luna observatio nova*," *Janus* 62 (1975): 13–15; Pierre Chabbert, "Pierre Borel

(1620?–1671)," *Revue d'histoire des sciences et de leurs applications* 21 (1968): 303–343.

19. For a discussion of the field of view of the early "Dutch" telescope, see John D. North, "Thomas Harriot and the First Observations of Sunspots," in *Thomas Harriot, Renaissance Scientist*, ed. John W. Shirley (Oxford: Clarendon Press, 1974), pp. 144–150, 158–160. When this instrument is adapted as a microscope, it has a similarly small field of view.

20. A. Van Helden, "The 'Astronomical Telescope,' 1611–1650," *Annali dell 'Istituto e Museo di Storia della Scienza di Firenze* 1 (1976): 13–35; Silvio A. Bedini, "Seventeenth Century Italian Compound Microscopes," *Physis* 5 (1963): 383–421.

21. On the focal lengths of objective lenses and the lengths of telescopes, see A. Van Helden, "The Telescope in the Seventeenth Century," *Isis* 65 (1974): 46–48.

22. Maurice Daumas, *Les instruments scientifiques aux XVIIe et XVIIIe siècles* (Paris: Presses Universitaires de France, 1953), pp. 47–49, 206–211. See also Olaf Pedersen, "Sagredo's Optical Researches," *Centaurus* 13 (1968): 139–150; Silvio A. Bedini, "The Makers of Galileo's Scientific Instruments," in *Atti del simposio internazionale di storia, metodologica, logica e filosofia della scienza, "Galileo nella storia e nella filosofia della scienza,"* 4 vols. (Florence: G. Barbera, 1967), 2 (part 5): 89–115.

23. A. Sachs, "Babylonian Observational Astronomy," in *The Place of Astronomy in the Ancient World, Philosophical Transactions of the Royal Society of London* series A, 276, no. 1257 (issued separately, London: Oxford University Press, 1974): 43–50.

24. *Ueber eine dioptra von Heron von Alexandria*, trans. Hermann Schöne, *Heronis Alexandrini opera quae supersunt omnia*, 5 vols. (Leipzig: Teubner, 1899–1914), 3:187–315; D. R. Dicks, "Ancient Astronomical Instruments," *Journal of the British Astronomical Association* 64 (1954): 77–85; Derek J. de Solla Price, "Precision Instruments: To 1500," in *A History of Technology*, ed. Charles Singer et al., 7 vols. (Oxford: Clarendon Press, 1954–1979), 3:582–619. See also, Alan E. Shapiro, "Archimedes's Measurement of the Sun's Apparent Diameter," *Journal for the History of Astronomy* 6 (1975): 75–83.

25. This is especially apparent in the case of the moon. See Olaf Pedersen, *A Survey of the Almagest* (Odense: Odense University Press, 1974), pp. 161–164.

26. Thomas S. Kuhn, *The Copernican Revolution: Planetary Astronomy in the Development of Western Thought* (Cambridge: Harvard University Press, 1957), pp. 124–133; Pedersen, *Survey*, pp. 19–22; Elizabeth L. Eisenstein, *The Printing Press as an Agent of Change: Communications and Cultural Transformations in Early Modern Europe* (New York: Cambridge University Press, 1979), 2:575–588; Ernst Zinner, *Leben und Wirken des Johannes Müller von Königsberg genannt Regiomontanus* (Munich: Beck, 1938); Donald DeB. Beaver, "Bernard Walther: Innovator in Astronomical Observation," *Journal for the History of Astronomy* 1 (1970): 39–43.

27. *De revolutionibus* (1543), dedicatory letter to Pope Paul III, trans.

Edward Rosen, *Nicolas Copernicus on the Revolutions, Nicholas Copernicus Complete Works* (London: MacMillan; Warszaw: Polish Scientific Publishers, 1972–), 2 (published separately, Baltimore: Johns Hopkins University Press, 1978): 4. Otto Neugebauer insists on a more accurate statement: "Since we shall find . . . that Copernicus preserved the equant also for Mercury we can now say that his aim was by no means to abolish the concept of equant, but exactly as his Islamic predecessors, to demonstrate that a secondary epicycle is capable of producing practically the same results (thanks to the smallness of the eccentricities) as Ptolemy's equant." ("On the Planetary Theory of Copernicus," *Vistas in Astronomy* [1968] 10:95.)

28. The title page of the first part of Tycho's most important work reads: *Astronomiae instauratae progymnasmata. Quorum haec prima pars de restitutione motuum solis et lunae stellarumque inerrantium tractat*, that is, "Preliminary exercise of restored (or renewed) astronomy, of which this first part deals with the restoration of the motions of the sun, moon, and fixed stars" (*Tychonis Brahe Dani opera omnia*, 15 vols. [Copenhagen, 1913–1929, reprint ed., Amsterdam: Swets & Zeitlinger, 1972] 2:3). See also, Robert S. Westman, "Three Responses to the Copernican Theory: Johannes Praetorius, Tycho Brahe, and Michael Maestlin," in *The Copernican Achievement*, ed. R. S. Westman (Berkeley, Los Angeles, London: University of California Press, 1975), pp. 306–308.

29. Victor E. Thoren, "New Light on Tycho's Instruments," *Journal for the History of Astronomy* 4 (1973): 25–45. The only technological difficulty cited by Thoren was that of making a five-foot hollow sphere (p. 30).

30. See note 24 above. Hero's dioptra was primarily a surveying instrument (see Price, "Precision Instruments," pp. 609–612). This section was written by A. G. Drachmann. Drachmann concludes: "Hero's dioptra remains unique, without past and without future: a fine but premature invention whose complexity exceeded the technical resources of its time" (p. 612).

31. John D. North, "The Astrolabe," *Scientific American* 230 (Jan. 1974): 96–106. North writes: "Imprecise as the astrolabe may have been in practice, it was undoubtedly useful, above all in judging the time" (p. 106). Willy Hartner, "The Principle and Use of the Astrolabe," in *A Survey of Persian Art from Prehistoric Times to the Present*, ed. Arthur U. Pope, 6 vols. (New York: Oxford University Press, 1938–1939), 3:2530–2554; reprinted in Willy Hartner, *Oriens-Occidens: Ausgewählte Schriften zur Wissenschafts- und Kulturgeschichte. Festschrift zum 60. Geburtstag* (Hildesheim: Georg Olms, 1968), pp. 287–311. Price, "Precision Instruments" (note 24), pp. 603–609. Levi ben Gerson started the chapter on the astrolabe in his *Astronomy* as follows: "There is great benefit to be gained from the use of the astrolabe because of the ease of using it, particularly for taking the altitude of a star at any place." See Bernard R. Goldstein, "Levi ben Gerson: On Instrumental Errors and the Transversal Scale," *Journal for the History of Astronomy* 8 (1977): 106.

32. Price points out that the primary purpose of the astrolabe was calculation; observation was a secondary function ("Precision Instruments," pp. 607–608). Ernst Zinner classified the astrolabe as a time measuring device (*Deutsche und Niederländische astronomische Instrumente des 11.–18. Jahr-*

hunderts, 2d ed., [Munich: Beck, 1967], pp. 135–145), and the torquetum as an educational device (pp. 177–183). Olaf Pedersen concludes on the basis of present information: "The absence of large instruments and observatories reveals that medieval astronomers were not generally interested in precision measurements or long-term programs of observation." See "Astronomy," in *Science in the Middle Ages*, ed. David C. Lindberg (Chicago: University of Chicago Press, 1978), p. 327. Richard of Wallingford introduced his *Tractatus rectanguli* as follows: "We designed the rectangulus to obviate the tedious and difficult work of making an armillary sphere" (*Richard of Wallingford: An Edition of his Writings with Introductions, English Translation and Commentary by J. D. North*, 3 vols. [Oxford: Clarendon Press, 1976], 1:407). See also Bernard R. Goldstein, "Theory and Observation in Medieval Astronomy," *Isis* 63 (1972): 39–47.

33. Tycho Brahe explained why the prognostications for the year 1588 prepared by two different mathematicians varied so much. One was based on the Alphonsine and the other on the Prutenic tables, which differed nineteen hours on the time of the vernal equinox, the starting point for all calculations of celestial aspects (J. L. E. Dreyer, *Tycho Brahe: A Picture of Scientific Life and Work in the Sixteenth Century* [Edinburgh: Black, 1890], p. 155).

34. Dreyer, *Tycho Brahe*, pp. 351–360, 387–389. Dreyer gives citations to historical investigations on this subject before 1890. G. L. Tupman, "A Comparison of Tycho Brahe's Meridian Observations of the Sun with Leverrier's Solar Tables," *The Observatory* 23 (1900): 132–135, 165–170. Thoren, "New Light on Tycho's Instruments" (note 29). Walter G. Wesley, "The Accuracy of Tycho Brahe's Instruments," *Journal for the History of Astronomy* 9 (1978): 42–53.

35. K. P. Moesgaard, "From Copernicus to Tycho Brahe, or from Blind Trusting to Frank Rejection of Ancient Records of Observation," in *Avant, avec, après Copernic: la représentation de l'universe et ses conséquences épistémologiques* (Paris: Blanchard, 1975), pp. 187–190; K. P. Moesgaard, "Copernican Influence on Tycho Brahe," *Studia Copernicana* 5 (1972): 31–55.

36. In observing the sun's declination, the apparent declination has to be corrected for the effects of refraction and parallax. Atmospheric refraction makes the sun appear higher above the horizon than it actually is, whereas parallax makes it appear lower. Tycho made the sun's refraction 34′ at the horizon and insensible above altitudes of 45°, while he accepted Ptolemy's maximum (horizontal) solar parallax of about 3′. Since his refraction correction was too small at higher altitudes and his parallax correction much too large, Tycho's corrected solar declinations were too great by several minutes of arc during the summer. Since the obliquity of the ecliptic is given by half the difference between summer and winter solstice meridian declinations of the sun, Tycho made the obliquity of the ecliptic about 2½′ too large. See Jean Dominique Cassini, "Les Élémens de l'Astronomie verifiez par Monsieur Cassini par le Rapport de ses Tables aux Observations de M. Richer faites en l'Isle de Cayenne," *Mémoires de l'Académie Royale des Sciences depuis 1666. jusqu'à 1699*, 11 vols. (Paris, 1729–1734) 8:55–59.

37. Hevelius described his instruments in *Machina coelestis pars prior* (Gdansk, 1673). For a comparison of the accuracies of Tycho and Hevelius, see J. B. J. Delambre, *Histoire de l'astronomie moderne*, 2 vols. (Paris, 1821; New York: Johnson Reprint, 1969), 2:476. Delambre concluded that Hevelius's observations were somewhat more accurate than those of Tycho. On the paucity of well-equipped observatories before 1670, see Olaf Pedersen, "Some Early European Observatories," *Vistas in Astronomy* 20 (1976): 24–25.

38. Eric G. Forbes, ed., *The Gresham Lectures of John Flamsteed* (London: Mansell, 1975), pp. 34–39. For Hevelius's objections to telescopic sights, see *The Correspondence of Henry Oldenburg*, ed. A. Rupert Hall and Marie Boas Hall (Madison: University of Wisconsin Press; London: Mansell, 1965–), 4:445–448, 5:181–182, 186–187, 241–242, 244. See also Eugene F. Mac-Pike, *Hevelius, Flamsteed and Halley: Three Contemporary Astronomers and Their Mutual Relations* (London: Taylor and Francis, 1937), pp. 1–16, 75–102.

39. Robert M. McKeon, "Les Débuts de l'Astronomie de Précision," Part I, "Histoire de la Réalisation du Micromètre Astronomique," *Physis* 13 (1971): 225–228, Part II, "Histoire de l'Acquisition des Instruments d'Astronomie et de Géodésie munis d'Appareils de Visée Optique," *Physis* 14 (1972): 221–242.

40. After having measured Mars's parallax during the opposition of 1671, Flamsteed concluded: "Haveing observed the distances & positions of the 3 stars by which Mars made his transit I find that Tycho erres 5 minute[s] at least both in the places & latitudes of them compared one wth another. and certeinely hee erres as much in many other[s]" (*Correspondence of Henry Oldenburg*, 9:327; *Phil. Trans.*, 1672, no. 89, p. 5119, 1673, no. 96, p. 6100 [misnumbered 6000]).

41. For the evolution of simple mechanical timekeepers from complex astronomical demonstration devices, see D. J. Price, *Gears from the Greeks, the Antikythera Mechanism, a Calendar Computer from ca. 80 B.C.*, *Transactions of the American Philosophical Society* 64, part 7 (published separately, New York: Science History, 1974); idem, "On the Origin of Clockwork, Perpetual Motion Devices and the Compass," paper 6 in *Contributions from the Museum of History and Technology*, United States National Museum Bulletin no. 218 (Washington, D.C.: Smithsonian Institution, 1959), pp. 81–112; idem, "Automata and the Origins of Mechanism and Mechanistic Philosophy," *Technology and Culture* 5 (1964): 9–23, reprinted in D. J. Price, *Science since Babylon*, 2d ed. (New Haven: Yale University Press, 1975), pp. 49–70.

42. Zdeněk Horský, "Astronomy and the Art of Clockmaking in the Fourteenth, Fifteenth and Sixteenth Centuries," *Vistas in Astronomy* 9 (1967): 25–34; H. Alan Lloyd, "Mechanical Timekeepers," in Singer, *History of Technology* (note 24), 3:648–661; H. von Bertele, "Precision Timekeeping in the Pre-Huygens Era," *Horological Journal* 95 (1953): 794–816.

43. Bernard Walther (1430–1504), student and patron of Regiomontanus, was the first astronomer to use a clock for astronomical measurements. He recorded its use in two observations, 1484 and 1487; Beaver, "Bernard Walther" (note 26). The clocks of Jost Bürgi were used by Landgrave Wilhelm IV of Cassel and his astronomer Christopher Rothmann of astro-

nomical observations (Bernhard Sticker, "Landgraf Wilhelm IV. und die Anfänge der modernen Astronomischen Messkunst," *Sudhoffs archiv* 40 [1956]: 15–25; Bruce T. Moran, "Princes, Machines and the Valuation of Precision in the Sixteenth Century," *Sudhoffs archiv* 61 [1977]: 209–228). The precise nature of Bürgi's contributions to precision timekeeping are assessed by von Bertele in "Precision Timekeeping" (note 42). Tycho Brahe described his clocks in *Astronomiae instauratae mechanica* (1598), *Opera omnia*, 5:29–30; see the translation by Hans Raeder, Elis Strömgren, and Bengt Strömgren, *Tycho Brahe's Description of his Instruments and Scientific Work, as Given in "Astronomiae instauratae mechanica"* (Copenhagen: Munksgaard, for the Danish Society of Science, 1946), pp. 29–30. For Tycho's reaction to the Landgrave's use of clocks, see his letter of 18 January 1587, *Opera omnia*, 6:68–69.

44. L. Defossez, *Les savants du xvii.e siècle et la mesure du temps* (Lausanne: Édition du Journal Suisse d'Horlogerie et de Bijouterie, 1946), pp. 242–243. For a convenient, rough graphic representation of the historical increase in accuracy of timekeepers, see Carlo M. Cipolla, *Clocks and Culture, 1300–1700* (New York: Walker, 1967), p. 59.

45. Henry C. King, *The History of the Telescope* (London: Charles Griffin, 1955), pp. 102–118; Allan Chapman, "Astronomia Practica: The Principal Instruments and Their Uses at the Royal Observatory," *Vistas in Astronomy* 20 (1976): 141–156. The increased *accuracy* of this new instrumentation and method was paralleled by a great increase in *convenience*. After 1700 the number of published observations increased enormously. See Dieter B. Herrmann, "Some Aspects of Positional Astronomy from Bradley to Bessel," *Vistas in Astronomy* 20 (1976): 183–186; idem, "Sternwartengründungen, Wissensproduktion und ökonomischer Fortschritt," *Die sterne* 51 (1975): 228–234.

46. Thoren, "New Light on Tycho's Instruments" (note 29).

47. The anchor escapement invented by William Clement in 1670 made possible long pendulums with very small amplitudes. The resulting long-case clock became the standard clock of observatories. The further improvement of the anchor escapement, the deadbeat escapement, invented by George Graham in 1715, demanded very high accuracy of machining, a limiting factor of general technology.

48. For the limitations of Greek mathematics, see Morris Kline, *Mathematical Thought from Ancient to Modern Times* (New York: Oxford University Press, 1972), pp. 173–176.

49. On the awkward compromise between Aristotle's cosmology and Ptolemy's mathematical hypotheses, see Edward Grant, "Cosmology," in Lindberg, *Science in the Middle Ages* (note 32), pp. 280–284, and Pedersen, "Astronomy," ibid., pp. 320–322. Copernicus characterizes the unsatisfactory nature of this compromise in his letter to Pope Paul III.

50. John E. Murdoch and Edith D. Sylla, "The Science of Motion," in Lindberg, *Science in the Middle Ages*, pp. 246–247.

51. The disintegration of the Aristotelian cosmos is the subject of Alexandre Koyré's *From the Closed World to the Infinite Universe* (Baltimore:

Johns Hopkins Press, 1957; New York: Harper & Brothers, 1958). See also Salomon Bochner, "Mathematical Background Space in Astronomy and Cosmology," *Vistas in Astronomy* 19 (1975): 133−161.

52. *Opera* (note 24), 1:5−7; *The Pneumatics of Hero of Alexandria*, ed. Bennet Woodcroft, trans. Joseph G. Greenwood (London, 1851), reprinted with introduction by M. B. Hall (London: MacDonald; New York: American Elsevier, 1971), p. 2.

53. *On the Natural Faculties*, bk. 1, chap. 13, C. G. Kuhn, ed., *Claudii Galeni opera omnia*, 20 vols. (Leipzig 1821−1833; reprinted, Hildesheim: Georg Olms, 1964−1965), 2:36−38; *Galen on the Natural Faculties*, trans. Arthur John Brock (London: Heinemann; New York: G. P. Putnam's Sons, 1916), pp. 59−61.

54. Edward Grant, ed., *A Source Book in Medieval Science* (Cambridge: Harvard University Press, 1974), pp. 435−551. See also Carl B. Boyer, *The Rainbow: From Myth to Mathematics* (New York: Thomas Yoseloff, 1959), pp. 110−123.

55. The name is variously rendered. I have followed Antonio Favaro (Galileo *Opere*, 20:316) and Stillman Drake (*Galileo at Work: His Scientific Biography* [Chicago: University of Chicago Press, 1978], p. 466. See Arturo Castiglioni, "Life and Work of Sanctorius," trans. Emilie Recht, *Medical Life* 38 (1931): 729−785, esp. 733−758; Ralph H. Major, "Santorio Santorio," *Annals of Medical History* 10 (1938): 369−381. For Santorio's share in the invention of the thermometer, see W. E. Knowles Middleton, *A History of the Thermometer and Its Use in Meteorology* (Baltimore: Johns Hopkins Press, 1966), pp. 5−14.

56. Middleton, *History of the Thermometer*, pp. 65−114.

57. *Astronomiae instauratae progymnasmata* (1602), *Opera omnia*, 2:64, 76−77, 136, 287.

58. *Astronomia pars optica* (1604), *Johannes Kepler Gesammelte Werke*, (Munich: Beck, 1937), 2:104−107, 117. Note that in the Rudolphine tables Kepler only printed Tycho's refraction tables, although he discussed his disagreement with them (*Gesammelte Werke*, 10:242−243, and p. 142 of the tables). See also, Gerd Buchdahl, "Methodological Aspects of Kepler's Theory of Refraction," *Studies in History and Philosophy of Science* 3 (1972): 265−298.

59. Koyré, *Galileo Studies*, trans. John Mepham (Atlantic Highlands, N.J.: Humanities Press, 1978), passim. See also, Koyré's "Galileo and the Scientific Revolution of the Seventeenth Century," *Philosophical Review* 52 (1943): 333−348, reprinted in Koyré, *Metaphysics and Measurement: Essays in Scientific Revolution*, ed. Michael A. Hoskin (London: Chapman and Hall; Cambridge: Harvard University Press, 1968), pp. 1−15; and his "Galileo and Plato," *Journal of the History of Ideas* 4 (1943): 400−428, reprinted in *Roots of Scientific Thought: A Cultural Perspective*, ed. P. Wiener and A. Noland (New York: Basic Books, 1957), pp. 147−175, and in Koyré, *Metaphysics and Measurement*, pp. 16−43.

60. "Galileo and the Scientific Revolution," in *Metaphysics and Measurement*, p. 13.

61. I. Bernard Cohen, *The Birth of a New Physics* (Garden City, N.J.: Doubleday, 1960), p. 106.

62. Norwood Russell Hanson, "Galileo's Discoveries in Dynamics," *Science* 147 (1965): 471.

63. See, e.g., E. J. Dijksterhuis, *The Mechanization of the World Picture*, trans. C. Dikshoorn (Oxford: Oxford University Press, 1961), p. 345; A. Rupert Hall, *The Scientific Revolution, 1500–1800: The Formation of the Modern Scientific Attitude* (London: Longmans, Green & Co., 1954), pp. 173–175, second edition (London: Longmans; Boston: Beacon Press, 1962), pp. 173–175; *idem., From Galileo to Newton, 1630–1720* (New York: Harper & Row, 1963), pp. 56–59; Richard S. Westfall, *The Construction of Modern Science: Mechanisms and Mechanics* (New York: John Wiley, 1971; Cambridge: Cambridge University Press, 1978), pp. 21–25.

64. E.g., "Galileo Gleanings – XXI: On the Probable Order of Galileo's Notes on Motion," *Physis* 14 (1972): 55–68. The fruits of this line of research have been summed up by Drake in *Galileo at Work* (note 55). See also Drake, *Galileo's Notes on Motion*, monograph no. 3, *Annali dell'Istituto e Museo di Storia della Scienza*, 1979.

65. Drake, "Galileo's Discovery of the Parabolic Trajectory," *Scientific American* 232 (March 1975): 102–110; "The Role of Music in Galileo's Experiments," *ibid.* 232 (June 1975): 98–104; *Galileo at Work*, pp. 84–90, 128–132.

66. Koyré, of course, reacted against what he felt was an overstressing of the role of experiment in Galileo's physics. Thus, Henry Crew and Alfonso de Salvio had translated Galileo's *comperio*, at the start of the Third Day of the *Discorsi* (1638), as "I have discovered by experiment" (*Dialogues Concerning Two New Sciences* [New York, 1914, reprinted, New York: Dover, 1954], p. 153). Koyré pointed out that the translators had been led by their epistemological predilections ("Traduttore-Traditore: A Propos de Copernic et de Galilée," *Isis* 34 [1943]: 209–210). Drake writes: "It is apparent that Galileo was describing [in his *Discorsi*] as a mental conception something he had carefully observed with his own eyes 30 years earlier. The first historians of science jumped to the conclusion that that was what he had done. Recent historians of science, critical of their predecessors, have jumped instead to the conclusion that Galileo worked from pure mathematics without empirical evidence; faith in ideal Platonic forms rather than attention to physical detail, they say, opened the way to modern science. As far as Galileo is concerned, the earlier historians came closer to the truth. What they lacked in philosophical insight they made up for in common sense." ("Galileo's Discovery of the Parabolic Trajectory," p. 110.)

67. Koyré, "An Experiment in Measurement," *Proceedings of the American Philosophical Society* 97(1953): 222–237, reprinted in Koyré, *Metaphysics and Measurement*, p. 94.

68. Koyré, *Metaphysics and Measurement*, pp. 98–108.

69. *Ibid.*, pp. 108–113. See also Pierre Costabel, "Isochronisme et Accélération, 1638–1687," *Archives internationales d'histoire des sciences* 28(1978): 3–20.

70. Cornelis de Waard, *L'expérience barométrique, ses antécédents et ses explications* (Paris: Thouars, 1936), pp. 101–117. W. E. Knowles Middleton, *The History of the Barometer* (Baltimore: Johns Hopkins Press, 1964), pp. 3–18.

71. De Waard, *L'expérience barometrique*, pp. 93–101; Middleton, *History of the Barometer*, pp. 8–10. See also Galileo's *Discorsi* (1638), in Stillman Drake, trans., *Galileo Galilei, Two New Sciences* (Madison: University of Wisconsin Press, 1974), pp. 24–26.

72. Middleton, *History of the Barometer*, p. 32.

73. Early Torricellian tubes had no scales. Descartes may have been the first to provide a scale. In 1647 he made two identical paper scales, one of which he sent to Mersenne so that their respective measurements could be compared (Middleton, *History of the Barometer*, p. 46). The results of Perier's "Puy de Dome" experiment were reported in two different sets of units (ibid., p. 52). For the variety of local units at this time, see William Barclay Parsons, *Engineers and Engineering in the Renaissance* (Baltimore: Williams and Wilkins, 1939; Cambridge, Mass.: M.I.T. Press, 1968), pp. 625–640.

74. The two different positions went back to antiquity, see De Waard, *L'expérience barométrique*, pp. 7–75. After Pascal's experiments, four positions were possible and adhered to by various thinkers. The peripatetics rejected the weight of the atmosphere and the vacuum above the mercury; Descartes accepted the weight of the atmosphere but rejected the vacuum; Giles Personne de Robervalle rejected the weight of the atmosphere but accepted the vacuum; and Pascal accepted the weight of the atmosphere and the vacuum above the mercury. See Dijksterhuis, *Mechanization* (note 63), pp. 444–445.

75. Reyer Hooykaas, *Religion and the Rise of Modern Science* (Edinburgh: Scottish Academic Press, 1972), pp. 54–74.

76. Paolo Rossi, *Francis Bacon, from Magic to Science*, trans. Sacha Rabinovitch (London: Routledge & Kegan Paul, 1968), pp. 26–27; for references to Bacon's statements on this subject in his various works, see p. 238, note 95. See also Rossi's *Philosophy, Technology and the Arts in the Early Modern Era*, trans. S. Attanasio (New York: Harper & Row, 1970), pp. 137–146.

77. Henry Power, *Experimental Philosophy* (London, 1664; reprinted New York: Johnson Reprint, 1966), pp. 192–193.

78. *Traitez de l'équilibre des liqueurs et de la pesanteur de la masse de l'air* (1663), I.H.B. Spiers and A.G.H. Spiers, trans. *The Physical Treatises of Pascal* (New York: Columbia University Press, 1937), passim. Boyle criticized Pascal for this vagueness and confusion of real experiments with thought experiments in *Hydrostatical Paradoxes* (1666), Thomas Birch, ed., *The Works of the Honourable Robert Boyle*, 6 vols. (London, 1744; 2d ed., London, 1772; reprinted, Hildesheim: Georg Olms, 1965–1966), 2:738–797.

79. Ibid., 1:12. The description of Boyle's air pump is on pp. 7–10 and Pl. I, facing p. 86.

80. Newton to Oldenburg, 6 February 1671/72, *The Correspondence of Isaac Newton*, 7 vols. (Cambridge: Cambridge University Press, 1959–1977) I. ed. H. W. Turbull: 96–97. Newton introduced the part of this paper dealing with the relationship between degrees of refrangibility and colors as

follows: "I shall now proceed to acquaint you with another more notable diffformity in its Rays, wherin the *Origin of Colours* is infolded. A naturalist would scearce expect to see ye science of those become mathematicall, & yet I dare affirm that there is as much certainty in it as in any other part of Opticks. [For what I shall tell concerning them is not an Hypothesis but most rigid consequence, not conjectured by barely inferring 'tis thus because not otherwise or because it satisfies all phaenomena (the Philosophers universall Topick,) but evinced by ye mediation of experiments concluding directly & without any suspicion of doubt.]" The part in square brackets was deleted from the version printed in *Phil. Trans.*, no. 80, (19 Feb. 1671/72), p. 3081. Note that Newton here argues that the conclusions of experimental science can be as certain as those of mathematical science.

81. For the objections of Hooke, Pardies, Huygens, and Linus, see Newton, *Correspondence* 1: passim. Newton reiterated his methodological position in his letter to Oldenburg of 6 July 1672 (*Correspondence*, 1:209).

82. It appears that Newton himself wished to construct a *mathematical* science of colors. See Johannes A. Lohne, "Newton's 'Proof' of the Sine Law and his Mathematical Principles of Colors," *Archive for History of Exact Sciences* 1 (1961): 389–405; Alan E. Shapiro, "Evaluation of Experiment and Newton's 'Achromatic' Dispersion Law," ibid. 21(1979): 91–128.

83. The relationship between scientists and craftsmen has usually been discussed in the context of the larger question concerning the relationship between science and technology in the Scientific Revolution. This is not the place to review this "internalist" versus "externalist" debate. I follow here the conclusions of A. Rupert Hall, presented in "The Scholar and the Crafts-man in the Scientific Revolution," in Marshall Clagett, ed., *Critical Problems in the History of Science* (Madison: University of Wisconsin Press, 1959), pp. 3–23.

84. Von Bertele, "Precision Timekeeping," (note 42).

85. I distinguish here between the *craftsman* (or artisan) and the *engineer*. The craftsman was a specialist who confined his work to a restricted range of products, clocks or lenses, for example; the engineer was a generalist who applied his expertise in construction and mechanisms to a wide range of unique problems.

86. This is, of course, a matter of perspective. Seen in comparison with the crafts of other civilizations over a long period of time, the European crafts were very progressive. In the short run, however, the sudden new demands of scientists presented problems the individual craftsmen often could not solve quickly. The guild structure tended to present barriers to innovation and its spread. A well-known example of this is the case of the Nuremberg copper worker Hans Spaichl who invented an improved slide-rest lathe in 1561. The local guilds tried strenuously to limit its use but were in the long run unable to do so (Friedrich Klemm, *A History of Western Technology* [German ed., 1954], trans. Dorothea Waley Singer [London: George Allen & Unwin, 1959; Cambridge, Mass.: M.I.T. Press, 1964], pp. 153–159). Moreover, the state of the art in individual crafts varied widely from place to place in Europe. For example, although England was more

dependent on coal mining than any other country in the seventeenth century, its mine-pumping technology was woefully backward compared with that practiced in central Europe (Graham Hollister-Short, "Leads and Lags in Late Seventeenth Century English Technology," *History of Technology* 1 [1976]: 159–183).

87. Van Helden, *The Invention of the Telescope* (note 6), pp. 16–25.

88. On glass quality see note 22. The limitations of the spectacle-making craft are reflected in contemporary accounts. Simon Marius could not obtain objective lenses of long focal lengths from the lens grinders in Nuremberg in 1608; see *Mundus jovialis* (1614), trans. A. O. Prickard, "The 'Mundus Jovialis' of Simon Marius," *The Observatory* 39 (1916): 370–371. Girolamo Sirturi traveled all over Europe in search for the secret of improving the telescope and found no craftsmen who could grind the lenses he required; see *Telescopium: sive ars perficiendi* (1618), pp. 22–30, trans. in Van Helden, *The Invention of the Telescope* (note 6), pp. 50–51.

89. In his early observations Thomas Harriot frequently mentioned Christopher Tooke, who made his telescopes. Very little is known about Tooke, who probably was not exclusively a lens grinder; see North, "Thomas Harriot" (note 19), pp. 141–142. It appears that, for the first twenty-odd years of the telescope's existence, no one in Europe could compete successfully with Galileo in producing research telescopes; see Drake, *Galileo at Work*, pp. 281–282, and Van Helden, "The Telescope in the Seventeenth Century" (note 21), p. 43, n. 28. Only around 1640 can we begin to find telescope and microscope makers who made contributions to the design and construction of these instruments. Thus, Francesco Fontana of Naples made popular the "astronomical" telescope (having a convex eyepiece) (Gino Arrighi, "Gli 'Occhiali' de Francesco Fontana in un Carteggio Inedito di Antonio Santini nella Collezione Galileiana della Biblioteca Nazionale di Firenze," *Physis* 6 [1964]: 432–448), and Johannes Wiesel of Augsburg invented the compound eyepiece (A. Van Helden, "The Development of Compound Eyepieces, 1640–1670," *Journal for the History of Astronomy* 8 [1977]: 26–37).

90. See note 89 above.

91. Newton *Correspondence*, 1:111–112, 154, 240, 251, 259, 271.

92. Ibid., pp. 73–81; *Phil. Trans.* no. 81, (25 March 1672), pp. 4004–4010.

93. Jean Picard described his instruments in *Mesure de la terre* (Paris, 1671), *Mémoires de l'Académie* (note 36), 7: Ar. V and IX. See also McKeon, "L'Acquisition des Instruments" (note 39), pp. 232–235.

94. See note 47 above.

95. Here instrument makers were hampered by more general technological restrictions. For example, the lack of large glass blanks of suitable quality restricted the apertures of refracting telescopes; the poor properties of speculum metal held back the development of the reflecting telescopes; the air pump was hampered by a lack of good seals; and advances in clockmaking depended on precision machining.

96. Only scientists who were very skillful with their hands, for example,

Galileo, Huygens, and Hooke, could make their own excellent instruments. Scientists rich enough to pay for the best instruments out of their own pocket, Hevelius and Boyle for instance, were very rare. The best collections of instruments were accumulated in Florence under the patronage of the Medicis and in Paris under the patronage of Louis XIV. Even though the English Crown established a Royal Observatory, no provisions for instruments were made, and Flamsteed had to rely on the patronage of wealthy individuals for his instruments.

97. When Christiaan Huygens solved the problem of Saturn's mysterious appearances he ascribed his success to superior telescopes. A number of men were offended by this assertion and Huygens was engaged in a long controversy by the Roman telescope maker Eustachio Divini concerning the respective qualities of their telescopes. See Van Helden, "Eustachio Divini versus Christiaan Huygens: A Reappraisal," *Physis* 12 (1970): 36–50.

98. *Opere*, 18:18.

99. Huygens, *De Saturni luna observatio nova* (1656), in *Oeuvres complètes de Christiaan Huygens*, 22 vols. (The Hague, 1888–1950), 15:172–177.

100. Van Helden, "Christopher Wren's *De corpore Saturni*," *Notes and Records of the Royal Society of London* 23 (1968): 219. See also Huygens, *Oeuvres complètes*, 3:419.

101. Cesi had founded the academy in 1603. It was his personal instrument of patronage, and therefore it disintegrated shortly after his death in 1630; see Drake, *Galileo at Work*, pp. 312, 444.

102. Richard F. Jones, *Ancients and Moderns: A Study of the Rise of the Scientific Movement in Seventeenth-Century England* (St. Louis: Washington University Studies, 1936, 2d ed., 1961), passim.

103. Hooke, *Micrographia* (London, 1665; reprint, New York: Dover, 1961; Brussels: Culture et Civilisation, 1966), preface, b2^v.

104. Rossi, *Philosophy, Technology and the Arts* (note 76), pp. 63–99.

105. See note 38 above.

106. Note 40. See also Francis Baily, ed., *An Account of the Revd. John Flamsteed* (London, 1835; reprinted, London: Dawsons of Pall Mall, 1966), p. 38.

107. Marcello Malpighi, *De formatione pulli in ovo* (London, 1673), in Howard B. Adelmann, *Marcello Malpighi and the Evolution of Embryology*, 5 vols. (Ithaca: Cornell University Press, 1966), 2:931–1013. See also Adelmann's commentary, pp. 817–930. Jan Swammerdam's treatise on bees, in which he shows that Aristotle's "King" was, in fact, a "Queen," that the drones are male and Aristotle's "females" neuter, was printed by Herman Boerhaave in *Biblia naturae*, 2 vols. (Leiden, 1737–1738). See Abraham Schierbeek, *Jan Swammerdam (12 February 1637–17 February 1680: His Life and Works* (Amsterdam: Swets & Zeitlinger, 1967), pp. 154–161.

108. Kepler did not apply his third law to Jupiter's satellites. This was first done by Gottfried Wendelin in a letter to Riccioli (*Almagestum novum* [note 14], 2:532). After giving his *regulae philosophandi* at the beginning of Bk. III of the *Principia*, Newton demonstrated how to derive universal gravitation from the phenomena. The first two phenomena adduced are

that the circumjovial and the circumsaturnian satellites obey Kepler's second and third laws.

109. C. Doris Hellman, *The Comet of 1577: Its Place in the History of Astronomy* (New York: Columbia University Press, 1944; reprinted, New York: AMS Press, 1971), pp. 118−233; idem, "The Role of Measurement in the Downfall of a System: Some Examples from Sixteenth Century Comet and Nova Observations," *Vistas in Astronomy* 9 (1967): 43−52.

110. Forbes, *Gresham Lectures* (note 38), p. 189. See also p. 169.

111. Boyle, *New Experiments Physico-Mechanical Touching the Spring of the Air* (1660), in Birch, *Works of Boyle* (note 78), 1:62−64. See also Marie Boas Hall, *Robert Boyle on Natural Philosophy: An essay with Selections from his Writings* (Bloomington: Indiana University Press, 1966), pp. 350−354. The fact that light passes through the space above the mercury in a Torricellian tube or through the evacuated receiver of an air pump was, of course, immediately apparent.

112. *Discorsi* (1638), in Drake, *Two New Sciences* (note 71), pp. 76−78. Boyle, *A Continuation of New Experiments Physico-Mechanical* (1669), in Birch, *Works of Boyle*, 3:256−259. Hall, *Boyle on Natural Philosophy*, pp. 347−349. Boyle could drop a feather in his receiver only from a height of twenty-two inches, but the experiment did show that all fluttering was eliminated.

113. Van Helden, "The Telescope in the Seventeenth Century" (note 21), pp. 46−48.

114. Cassini determined the rotation periods of Mars and Jupiter and discovered four satellites of Saturn as well as the division in Saturn's ring. For his achievements, see Delambre, *Histoire de l'astronomie moderne* (note 37), 2:686−804.

115. The new map of France presented by Picard and La Hire in 1684 moved the point of Brittany 1½° of longitude to the east and the south coast of France ½° of latitude to the north, compared with previous maps. See G. R. Crone, *Maps and Their Makers: An Introduction to the History of Cartography* (London: Hutchinson, 1953), pp. 128−130.

116. In 1672−1673 Richer measured meridian altitudes of the sun at Cayenne, near the equator, where because of the sun's high altitudes atmospheric refraction is minimal. His observations showed that the obliquity of the ecliptic is 2′ less than Tycho had made it. This dovetailed with a solar parallax of not more than 10″ (instead of Tycho's 3′), confirmed by the parallax measurements of Mars made that year. See Cassini, "Les élémens de l'Astronomie" (note 36), pp. 55−117.

117. Eric G. Forbes, A. J. Meadows, and Derek Howse, *Greenwich Observatory*, 3 vols. (London: Taylor and Francis, 1975), 1, *Origins and Early History (1675−1835)*, ed. Eric G. Forbes: 91−107.

IV

ROBERT HOOKE,
MECHANICAL TECHNOLOGY, AND
SCIENTIFIC INVESTIGATION

Richard S. Westfall

Like seventeenth-century science, which aspired to be not only true but fruitful, scholarship related to seventeenth-century science has been fertility itself in the production of works concerned with the two principal elements of my title, technology and scientific investigation. It is small wonder. No single factor has contributed more to making the world what it is today, for better or for worse, than the union of modern science and technology. Look where you will—at the kitchen, where a hundred appliances have utterly transformed an institution as old as man; at our transportation system, which encourages and permits a symposium on science, technology, and society in postrevolutionary England to be assembled from the corners of the world; at military techniques, which are too hideous in their destructive potential to bear contemplation—scientific technology has transformed every facet of our lives to the degree that we cannot seriously imagine what they would be without it. As it is almost universally agreed that the seventeenth century witnessed the birth of modern science, what could be more reasonable than to look there as well for the origins of modern scientific technology?

Nor is the seventeenth century wanting in materials relevant to such a search. At its very dawn the echoing voice of the Lord Chancellor, Francis Bacon, proclaimed a new purpose for natural philosophy, the establishment of the kingdom of man. "The world was made for man, Hunt," Bacon told his servant, "not man for the world." The admonition to his servant repeated one of the central themes of his philosophy. "Human knowledge and human power meet in one," he declared in the *Novum Organum*; "for where the cause is not known the effect cannot be produced. Nature to be commanded must be obeyed; and that which in contemplation is as the cause is in operation as the rule."[1] To this end he directed his new method. Where Aristotelian logic aimed, he said, at the invention of arguments, his method pursued the invention of arts. Where Aristotelian science aimed at victory in disputation, his method sought "to command nature in action."[2] Sometimes the rubrics in which Bacon expressed his vision ring strange to the twentieth-century ear—"the power and dominion of the human race . . . over the universe," "the empire of man over things," "that right over nature which belongs to [the human race] by divine bequest."[3] His much repeated phrase "the kingdom of man" expressed his goal that science would restore man to the dominion over nature lost in the Fall. Not the least of the changes wrought by nearly four hundred years is the virtual destruction of Bacon's theological idiom. We now prefer to say in wholly secular terms that knowledge is power; and we know today, to an extent that Bacon never dreamed, how vast the power of knowledge is. So far has modern science and the technology it nurtured outstripped the vision of its prophets. For all that, Benjamin Farrington has labeled Bacon "the philosopher of industrial science."[4]

Baconian utilitarianism has furnished the argument central to much of the literature concerned, immediately or tangentially, with the seventeenth-century sources of scientific technology. R. F. Jones's *Ancients and Moderns* did not address itself to technology as such but rather considered the impact of Baconian utilitarianism on the Puritan consciousness, especially at the time of the Civil War.[5] In *Science, Technology, and*

Society in Seventeenth Century England, R. K. Merton picked up the Puritan theme, combined it with Weber's thesis, and argued that Baconian utilitarianism furnished the cultural validation that promoted large-scale pursuit of science and technological innovation.[6] More recently, similar arguments have been pursued at length by Christopher Hill and Charles Webster.[7] Webster, who has emphasized the Baconian role in his story by naming his book *The Great Instauration,* urges that we find the source of modern science and especially modern technology in the Puritan dream that the completed Reformation would bring in its train a revival of learning. The fruits of experimental science would manifest themselves in the reconstruction of society.

Baconian utilitarianism has received unexpected illumination from another source, the growing interest in the hermetic tradition. We have learned that the intent to manipulate nature for useful ends was a constant presence in that tradition and have recognized with a start that many of those Puritans most outspoken in the pursuit of useful knowledge had Paracelsian chemistry in mind, an enterprise more closely affiliated with the hermetic tradition than with those achievements that made seventeenth-century science a true revolution in Western thought.

Having achieved these insights, we have also found that many of the familiar passages in Bacon hold levels of meaning we scarcely suspected. "The End of our Foundation," the member of Salomon's House announced, "is the knowledge of Causes, and secret motions of things, and the enlarging the bounds of Human Empire, to the effecting of all things possible."[8] The effecting of all things possible—the goal of Baconian utilitarianism, as we examine it in Bacon's own words, partakes more of the alchemist's dream than of humdrum engineering. It was not wholly by accident that the example of operative knowledge which came most readily to his mind was the making of gold. One of the passages in his works most revealing of his purpose is aphorism 50 in the second book of the *Novum Organum,* on "Polychrest Instances, or Instances of General Use."[9] It suggests that when Bacon spoke of useful knowledge, he thought of manipulating the powers and active virtues of matter in order to transform bodies and create new substances. Salomon's House, in the *New Atlantis*, had deep mines in which artificial metals were produced by burying prepared materials under conditions similar to those by which nature was thought to generate metals. By varying and mixing earths, they produced new plants and turned one into another. With animals they did much the same. "We make a number of kinds of serpents, worms, flies, fishes, of putrefaction; whereof some are advanced (in effect) to be perfect creatures, like beasts or birds, and have sexes and do propagate. Neither do we this by chance, but we know beforehand of what matter and commixture what kind of those creatures will arise."[10]

Bacon argued that natural philosophy is divided into two branches, speculative and operative, "the one searching into the bowels of nature, the other shaping nature as on an anvil."[11] Shaping nature on an anvil meant "profound and radical operations on nature," "profound alterations made in bodies," "remarkable transformations and alterations of bodies."[12] His *Sylva Sylvarum*, he said, was not a natural history but "a high kind of natural magic. For it is not a description only of nature, but a breaking of nature into great and strange works."[13] Knowledge that is confined to efficient and material causes does not touch the deeper boundary of things. "But whosoever is acquainted with Forms, embraces the unity of nature in substances the most unlike; and is able therefore to detect and bring to light things never yet done, and such as neither the vicissitudes of nature, nor industry in experimenting, nor accident itself, would ever have brought into act, and which would never have occurred to the thought of man."[14]

A strange philosopher of industrial science: rather let us call him a philosopher of Faustian science, for he had more in common with Faust than with Ford. Let us also use the ambiguities in Bacon as a warning not to conclude too readily that the familiar theme of Baconian utilitarianism must also have entailed the presence of scientific technology. That the theme was pervasive throughout the English scientific community no one doubts, but with regard to technology we still lack detailed studies of specific innovations in relation to the scientific principles supposed to have fostered them. In his massive study of utilitarianism among the Puritan reformers, Webster concluded with five case studies of utilitarian projects—such things as machine coinage and the Down survey of Ireland. I do not find among them anything to which the word *science*, as I understand it, properly applies. Perhaps it is revealing that a work devoted to demonstrating the utilitarian dimension of seventeenth-century science found only one technological innovation it could attribute to the Puritan scientists—a new dye; and it offered no evidence that the dye passed into general use. It is against this background that I propose to comment on Robert Hooke, mechanical technology, and scientific investigation. Though I am under no illusion that the case of one man can settle the broad issues involved, it can offer empirical evidence relevant to two important and controversial historical questions, on the sources of scientific technology and the origins of modern science.

A better subject for careful scrutiny in this respect could scarcely be imagined. Robert Hooke was a scientist of major importance, though of how major importance has been a matter for debate over the years. Hooke has tended to elicit the devotion of partisan supporters in a way that few other scientists have. I believe this has stemmed from the combined effect of his manifest ability in a wide number of fields and the injustices, or apparent injustices, he sustained from a number of men, especially Isaac Newton. Despite his partisans, I think there is no informed historian of science today who would place Hooke in the highest category, in the company of scientists such as Kepler, Galileo, and Newton. The adjective *major* is broad enough, however, that I will nevertheless apply it to Hooke as scientist. He participated in the experiments that led to Boyle's Law—participated to the extent that a number of scholars have wondered if the law, which is typical of Hooke's scientific bent but not of Boyle's, should not bear another name. And whatever his contribution to Boyle's Law, there is no doubt that Hooke generalized it to the law of elasticity that does bear his name. He was the first to correct the misunderstanding of the mechanics of circular motion, which before him was seen as an equilibrium of opposed forces, and to formulate the program in planetary dynamics that Newton carried out in the *Principia*. In optics Hooke called attention to the color phenomena of thin plates and proposed that they were periodic, thereby stimulating Newton to the investigation, ultimately published as book two of the *Opticks*, which first demonstrated the periodicity of an optical phenomenon. He was one among several associated with the Royal Society who used the air pump to investigate the relation of air to combustion and animal life. Some have argued that they effectively discovered oxygen a century before Priestley and Lavoisier. Most historians find this judgment exaggerated; nevertheless, Hooke and others did significantly extend previous understanding of these matters. His study of fossils, which he adamantly refused to believe were anything but the remains of once living creatures, whether or not anything like them still existed on the earth, made him one of the leading figures in the still inchoate science now called geology. Above all, his *Micrographia*, the first great publication of microscopic observations, opened a new world to mankind, inaugurated the study of insect anatomy, and coined the biological usage of the word *cell*. Add it all up and Hooke's contribution to science does not, it is true, begin to equal a *Principia*. On the other hand, more than 99.9 percent of all scientists who have ever lived would have been glad to settle for an achievement of equal permanence, and so I feel comfortable enough with my assertion that Hooke was a scientist of major importance.

He was also an inventor of major importance. About this aspect of his activity there is, I believe, no debate. With a significant choice of words, Lalande called him "the Newton of mechanics." Consider briefly a list of his accomplishments. Hooke invented the universal joint. He added, in one way or another, to every important scientific instrument that appeared during the century. He improved microscopy. To the telescope he contributed the cross-hair sight, the iris diaphragm, and the worm gear as a device to divide small units of arc with precision. The air pump in its enduring form was his work. Hooke has been called the father of scientific meteorology. He devised the first wheel barometer on which the rotation of a needle instead of the height of a column of mercury registered the pressure. He proposed that the temperature at which water freezes furnish the zero point of thermometers, and he invented an instrument to calibrate thermometers. He conceived a weather clock which would record barometric pressure, temperature, rainfall, humidity, and wind velocity on a rotating drum.

Most important of all perhaps, though shrouded in obscurity, apparently forever, was his contribution to the spring-driven watch. Inability to determine longitude was the major deficiency of navigation at the time. A portable precision chronometer was an obvious solution to the problem, indeed the solution finally perfected in the eighteenth century. Hooke reasoned that he could make the desired instrument by "applying Springs to the Arbor of the Ballance of a Watch, for the regulating the vibrations thereof in all postures."[15] That is, he would attach a spring to the balance wheel to control its oscillations as the pendulum controls the motion of a clock. It is impossible to learn the exact nature of what Hooke did. By 1660 he had advanced far enough to enter into negotiations with three men (among them Robert Boyle) who were to back the invention financially, though at the last moment Hooke withdrew because he did not like the terms of the patent. When Christiaan Huygens constructed a watch in 1674 with a spiral spring controlling the balance, Hooke accused Henry Oldenburg of peddling his invention abroad.[16] He had the clock maker Thomas Tompion make such a watch for presentation to Charles II, and on it he engraved his claim: "Robert Hooke inven. 1658. T. Tompion fecit 1675." Unfortunately there is no evidence that he had used a spiral spring in his earlier watch or indeed that he had made a working model at that time. For my purposes, however, it is not necessary to probe this riddle further.* Knowledgeable people agree unanimously that Hooke made important contributions to chronometry.

The mere recitation of Hooke's accomplishments, though imposing enough, loses the full flavor of his inventiveness. To savor that we need to listen to Hooke himself commenting on Hevelius's mistake in not using telescopic sights. Even when we discount many of the particulars as the exaggerations of vanity, it provides an impressive view of inventive genius overflowing with a torrent of new ideas. Hevelius's instruments

*See pp. 37–42 for Marie Boas Hall's description of and comments on the Hooke-Oldenburg controversy. (Editor's note.)

were good, Hooke agreed, but he did not want the interested public to think they represented the ultimate.

For I can assure them, that I have my self thought of, and in small modules try'd some scores of ways, for perfecting Instruments for taking of Angles, Distances, Altitudes, Levels, and the like, very convenient and manageable, all of which may be used at Land, and some at Sea, and could describe 2 or 3 hundred sorts, each of which would be every whit as accurate as the largest of Hevelius here described, and some of them 40, 50, nay 60 times more accurate, and yet everyone differing one from another in some or other circumstantial or essential part. And that this may not seem altogether so strange, I will assure them that I have contrived above 20 ways for dividing the Instrument, each of them as much distinct from each other as this of Hevelius, and that of Diagonals, and yet every one capable of as great certainty and exactness at least, and some of them 100 times more. I have above a dozen several ways of adjusting the Perpendicularity or Horizon-tality of Instruments, all as exact as the common Perpendicular, and some of them very much more, even to what accurateness shall be desired, and yet each of these very differing one from another. I have as many differing kinds of Sights, for improving, directing, adjusting and ascertaining the Sight, some of which are applicable to some particular uses, but some for all, by means of which that part also may be improved to what accurateness is desired.[17]

For all his ingenuity, Hooke was not among the most articu-late advocates of Baconian utilitarianism. He saw himself first as a natural philosopher, and he worried that excessive con-cern for immediate use might distort the philosophic enter-prise. He compared natural philosophy to a growing plant. The design in organizing the Royal Society had been to promote the growth of natural philosophy " 'till it attain to a State of Perfection of Flowering and Blooming, and of pro-ducing Seed, and becoming Fruitful." Savoring the analogy, he pursued it at some length. He described how a seed sends out fibrils into the soil, by which it draws nutriments, grows, and produces leaves and branches which draw from the air enlivening juices which in turn feed the plant and roots, allow-ing further growth until the plant reaches full maturity.

Nor will a skillful Gardener suffer it to be tapp'd to have the nutritive and vital Juice drawn off to be used for other purposes than the nutrition of its own Body; well knowing it would hinder its Growth, prolong the time of its coming to Maturity, weaken its Constitution at least, if not render it wholly barren.

In a similar way natural philosophy receives from the phenomena of nature its first information, which enables it to
branch out into conclusions which then strengthen further
inquiries, until by natural reinforcements it arrives at a perfect
understanding of nature; "nor must the Natural Course of
Circulation be stopp'd or diverted, 'till the utmost Perfection
be attain'd, if at least it be aim'd to compleat and make it
prolifique."[18]

Despite its reservations about the rush for immediate utility,
the comparison assumed that the final goal of natural philosophy was fruitfulness, what Bacon called the comfort and
convenience of mankind. In other passages Hooke made it
clear that he did subscribe to this ultimate goal, shorn of its
hermetic overtones in his more modest statement of it and
hence approximating the outlook of modern technology. The
business of natural philosophy, he asserted in the introductory
paragraph of his "General Scheme, or Idea of the Present
State of Natural Philosophy," is to arrive at a perfect knowledge of nature, not just for itself, "but in order to the inabling a
Man to understand how, by the joyning of fit Agents to Patients according to the Orders, Laws, Times and Methods of
Nature, he may be able to produce and bring to pass such
Effects, as may very much conduce to his well being in this
World, both for *satisfying his Desires*, and relieving of his
Necessities."[19]

Let me add one further element to the picture. Whenever
Hooke came down from the high generality of Baconian utilitarianism to consider specific technological problems—and
this was very frequently indeed—he approached them almost
invariably from the side of science. His method of dividing a
degree by the worm gear was, he declared, "so Mathematically
and Mechanically true, that 'tis hardly to be equalized by any
other way of proceeding."[20] He proposed a new windmill in
which the vanes were so designed that "the Force of the Wind
becomes equal upon every Part of the Vane, from the Center
to the Tip, or Extremity thereof. An equal Progression of
Wind causing every Point of the whole Vane to make an equal
Arch of Rotation, or an equal Angle at the Axis."[21] He argued
that his instrument to measure the strength or velocity of the

wind would make it possible truly to determine with what wind, what manner of trimming the sails, and what burden a ship makes the best way. "This is a Proposition hitherto only prov'd by guess and strong Opinion, for the most part very prejudicate and precarious; but by this means it might be brought to a certain Standard of measure. . . ."[22] In the attempt to reduce various technical problems that had been attacked over the centuries by trial and error to "a certain Standard of measure," surely we stand face to face with the ideal of scientific technology.

The ideal of scientific technology is, of course, only a restatement of Baconian utilitarianism. The problem I propose is to determine not whether men in the seventeenth century subscribed to the ideal but whether the ideal was translated into fact; for the purpose Robert Hooke, one of the leaders of the English scientific community during the Restoration and the leading inventor of his age, appears to be a perfect subject to examine. Among his works are three examples that supply the details of his scientific analysis of technical problems. By looking at his demonstrations we can arrive at a clearer understanding of the relations of science and technology in the late seventeenth century.

In his Cutlerian lecture *Lampas* Hooke described a number of lamps designed to burn with a steady, uniform flame. To understand his point we must realize that he meant to use the lamps as sources of heat rather than light. We can compare them with two radically different devices. On the one hand, the lamp was Hooke's Bunsen burner in which he could shape glass and heat reactants in chemical experiments. On the other hand, it was a thermostatically controlled space heater to keep plants warm enough to grow and to hatch eggs artificially. To be honest, I do not find that any of the uses he mentioned, with the possible exception of chemical experimentation, necessitates the degree of ingenuity he expended. That is Hooke's problem, however, not mine, and reading *Lampas* one is not likely to miss Hooke's delight in spinning off device after device—even unto the numbers he cited to Hevelius—by which to maintain a steady flame in a lamp.

Hooke insisted that what he proposed about lamps was a consequence of his hypothesis of combustion, presented a decade earlier in *Micrographia*. By understanding the "nature and causes" of fire, he argued, one can discover the inconveniences that may occur in lamps and "more readily and scientifically find a cure and prevention of those inconveniences which he that is ignorant of can but hoodwinked grope after, and at best can but hope possibly after long puzling himself in vain attempts and blind trials, nothing to the purpose, he may at length stumble upon that which had he been inlightned by the true Theory, he would have readily gone to at the first glance."[23] Yet when we examine the scientific foundation of this invention, what do we find? Hooke described his hypothesis of combustion in some detail as a dissolution of sulfurious matter by nitrous salts in the air, with concomitant generation of heat. It was less a hypothesis than an analogy to the dissolution of substances in acids; but he also forgot along the way that he had introduced it as a hypothesis and concluded by calling it a theory. He described as though it were empirical fact how particles of fuel, vaporized by heat, fly off the wick of a lamp and ascend in the air, where they are dissolved; he even ascribed parabolic trajectories to them. The net result of this ponderous deployment of theory, nothing of which survived for long, was the conclusion that lamps need a steady supply of fuel to burn steadily. Surely any blacksmith could have told him as much. If this is the fruit of scientific technology, it is disappointingly meager.

Providing a steady flow of fuel posed a second problem, which he also attacked with science. Hooke claimed it started him inquiring into a counterpoise for liquids—a subject of great use in hydraulics. In his usual fashion, he described a number of possible counterpoises. Some were ingenious; some were simple and obvious and, he admitted, already commonly used. He clearly found the first one of primary interest; in the lecture he devoted the most attention to it. The lamp was a spherical vessel with the wick along the equator at the surface of the fuel, which initially filled exactly one-half of the sphere. Above the fluid was a hemispherical float or counterpoise turning on a horizontal axis so that it descended on the side opposite the wick. Hooke was easily able to demonstrate that for any position of the counterpoise, when one ignored the mathematically distinguishable segments of fuel and float that were poised symmetrically in relation to a vertical line through the axis, the remaining part of the fuel was in equilibrium with a segment of the counterpoise which was twice as large (see fig. 1). Hence, if the counterpoise were made of a substance with a specific gravity exactly half that of the fuel, the counterpoise would descend uniformly as the lamp burned and would always maintain the fuel next to the wick at a constant level.

Let us examine this demonstration. It consistently used the word *weight* where the word (or the concept) *moment* was needed. There is no derogation of Hooke implied in this statement. The science of mechanics was still sorting out the elements of circular motion. Newton, who did much to clarify them, made at least one egregious blunder in a similar problem in the first edition of the *Principia*.[24] Our task, however, is not to compare Hooke to others of his age but to judge whether the devices he proposed on the basis of scientific analysis would work. In this case Hooke was saved by the

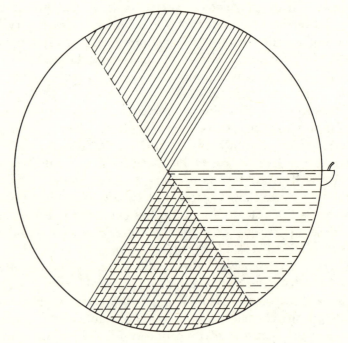

Figure 1

symmetry of the circle, and it is only fair to add that he insisted on the need for a vessel of circular cross-section. Nevertheless, extended to other than circular vessels, the inherent defects of the analysis could easily have led to mechanical impossibilities. And the problems did not end there. As Hooke remarked, to find a material of exactly half the specific gravity of the fuel would be difficult—we might rather be inclined to say impossible in that age. Here the defects of his analysis did lead to a mechanical impossibility. Again he used the word *weight*, this time in place of *specific gravity* (or its equivalent), and misled by the word, said that one could make a hollow float with a weight suspended somewhere on the line that bisected the hemisphere such that the total weight of the counterpoise was half that of the fuel.[25] Obviously this would not have worked. A counterpoise made in this manner would have been in the equilibrium Hooke desired only in two positions—when the lamp was completely full before it was lit, and when it was completely empty after the flame had gone out. The show of scientific analysis was an illusion covering a misconception that flowed from the current level of the science of mechanics.

A second example of Hooke's scientific analysis of technical problems concerns ships. It differs from the first example in involving, rather than a new invention, an attempt to determine the ideal way to trim a sail. Navigators and shipwrights were almost unanimous, Hooke stated, in assuring him that bellying sails were most effective in driving a ship. By considering "the Nature and Power of the Wind upon the Area of a Sail," however, he arrived at what he called "the evident Demonstration" of the opposite. The "evident demonstration" treated the issue as a problem of impact. It was hampered by two irremediable defects: its concept of the force of a body in impact had been rendered obsolete by Huygens and Newton; and in any case, impact eventually proved to be a mistaken approach to such problems. Bodies in motion, Hooke argued, impress on others they strike motions that are in proportion to the gravities and the velocities of the bodies that strike. Suppose that water is 900 times denser than air. If a quantity of water moves a body, thirty times as much air with a velocity thirty times greater will move it equally. Consider a sail of some given height and breadth *ab* set perpendicular to the wind. Let *abcd* represent the total volume of air that moves against the sail in a given time. Now let *abon* represent a prism of water on the same base such that *na*, the length of the prism of water, equals *da*—the length of the prism of air—divided by thirty. The prism of water moves in the opposite direction against the sail (see fig. 2). Hooke's analysis involved the unfortunate consequence that the ship must sail right side up and upside down at the same time, a difficult posture in the best of circumstances, and one perilously close to the vulgar notion of shipwreck.

d   a n

c   b o

Figure 2

I say [he continued] the Sail shall not be mov'd either way, but remain in an equilibrium: For as the Velocity of the motion of Water an, 1 is to the Velocity of the Air ad 30, so the Gravity of the Prism of Air abcd 30, to the Gravity of the Prism of Water abon 900. Now because the same power is imprest on the Sail, whither the Cylinder of Water be mov'd against the Sail from no to ab, or the Sail be mov'd against the Water from ab to no; if the said Cylinder of Air be made one degree swifter, it must drive the Sail from ab, to no.

If it matters much after this beginning, Hooke was unable to extract an answer to his question with the sail set perpendicular and the ship running before the wind. When it was on a tack, however, he took the component perpendicular to the sail as the force of the wind driving the ship, and he argued that with a bellying sail its shape is not symmetrical on a tack, and the various components perpendicular to elements of the sail must hamper each other's effect. Hence a ship receives more force from a taut sail.[26] Hooke complained that practical seamen would not listen to his demonstration. Perhaps we can appreciate their preference for knowledge gained from experience at sea. We can explain away the worst features of Hooke's demonstration by the fact that he was an old man in decline when he produced it. The fact remains, however, that the propositions on the resistance of fluids, the most analogous parts of Newton's *Principia*, were also its weakest section. Science did not then have in its power the tools to answer correctly the question Hooke posed.

The third example is Hooke's object of pride, the watch. Though, as I have indicated, it is impossible to establish the exact extent of Hooke's contribution to the watch, for my purposes it does not matter. Let us assume that Hooke applied a spiral spring to the balance as he said he did. The question we need to answer is what relation the resulting device had to scientific theory. In the work that announced Hooke's Law he appeared to grasp intuitively that a vibrating spring is dynamically equivalent to a pendulum. Intuition is an unsatisfactory foundation for scientific technology, however. For that we need full understanding of the general principles in play. Fortunately for our investigation, Hooke undertook demonstrations of the dynamics both of pendulums and of vibrating springs.

He discussed pendular motion before the Royal Society in 1666. It presents two problems, he told them, the determination of velocity in each vibration and the length of periods for different arcs. Velocity derives from the proportion between "the quantity of strength, and the bulk of the body to be moved." The bulk of the pendulum's bob remains constant. Gravity supplies the strength, its effective component increasing with displacement. Hence the motion is swifter in larger arcs. If the pendulum moves in a certain curve, which is nearly a circle, a curve determined by "the proportion of the length of the intercepted arches to the length of the perpendicular lines of attraction terminating those arches," all swings will be isochronous. Specifically, the isochronous curve demands that AB^2 (where AB is the length of the arc) be proportional to BC (the vertical height of the end of the arc) (see fig. 3).[27] What are we to make of this "demonstration"? It appears to me that Hooke started with the property of circles that the versed sine (or BC) varies as the square of the chord and reasoned that there must be a curve that is nearly a circle for which the versed sine varies as the square of the arc. From Galileo he knew the equivalent of the equation $v^2 = 2as$. He appears to have con-

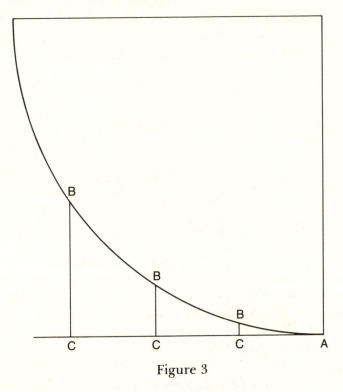

Figure 3

verted it to a dynamic formula $v^2 \propto Fs$, or better $v^2 \propto F$, where the force F would equal BC, the vertical distance through which gravity operates. Boldly (or perhaps blindly) neglecting the complications of nonuniform acceleration, he reasoned that the time for any motion is constant when the velocity increases in proportion to the distance. Both the distance squared (AB^2) and the velocity squared vary as BC. Ergo, etc. Let us not be so churlish as to inquire what he could have meant by the velocity of a body that accelerates nonuniformly from rest to its maximum velocity.

Force is proportional to velocity squared—Hooke made this formula the foundation of all his efforts in mechanics. Early in 1669 he performed an experiment before the Royal Society to prove that the "strength" of a body in motion is in duplicate proportion to its velocity. The experiment failed, but a week later he tried two more experiments to show that the "force in moving bodies" is in duplicate proportion to the velocity "so that there is required a quadruple weight to double the velocity."[28] In his *Lampas* he stated what he called a "General Rule of Mechanicks."

Which is, that the proportion of the strength of power of moving any Body is always in duplicate proportion of the Velocity it receives from it [*sic*]; that is, if any Body whatsoever be moved with one degree of Velocity, by a determinate quantity of strength, that body will require four times that strength to be moved twice as fast, and nine times the strength to be moved thrice as fast, and sixteen times the strength to be moved four times as fast, and so forwards.[29]

The rule holds good, he asserted, in the motion of bullets shot from cannons, wind guns, and crossbows, in the motion of arrows shot from bows, of stones thrown by hand or by sling, of pendulums moved by gravity, of musical strings, of springs and vibrating bodies in general, of wheels and flies turned by weights or springs, of falling bodies—in a word, it is a general rule, as he said, which holds for all mechanical motions.

Manifestly, Hooke's rule covered considerable vagueness and confusion, and the vagueness and confusion were not his alone. He reflected all the problems of the age as it tried to come to grips with the concept of force. The confusion expressed itself in the multiplicity of terms. Hooke spoke of "strength," "quantity of strength," "force," "force of a moving body," "pressure," "power," and he seemed to use the words and phrases interchangeably. Once, determined apparently to leave nothing to chance, he referred to "force, pressure, indeavour, impetus, strength, gravity, power, motion, or whatever else you will call it."[30] Behind the multiplicity of terms lay a problem of concepts. Most of the time what Hooke called force was equivalent to our *work*, but sometimes it was identical to what we now call force. It does not matter that others were equally muddled until Newton began to sort the concepts out. In technology, the principle invoked must lead to a machine that works. The product of confusion is not technology; it is a bungle. Caught up in this vagueness, Hooke's dynamics was incapable of offering the necessary clarity.

With his dynamic concepts in mind, let us look at Hooke's discussion of vibrating springs, one of the early attempts to analyze what we now call simple harmonic motion. He explicitly included spiral springs among those covered. His treatment rested directly on his own law of elasticity, that the "power" of a spring is directly proportional to its degree of flexure. To each of the infinite degrees of strain there corresponds a specific intensity of power or force. "And consequently all those powers beginning from nought, and ending at the last degree of tension or bending, added together into one sum, or aggregate, will be in duplicate proportion to the space bended or degree of flexure. . . ." Thus, if the sum of all the powers corresponding to a strain of one equals one, it equals four when the strain is two.

The spring therefore in returning from any degree of flexure, to which it hath been bent by any power receiveth at every point of the space returned an impulse equal to the power of the Spring in that point of Tension, and in returning the whole it receiveth the whole aggregate of all the forces belonging to the greatest degree of that Tension from which it returned; so a Spring bent two spaces in its return receiveth four degrees of impulse, that is, three in the first space returning, and one in the second; so bent three spaces it receiveth in its whole return nine degrees of impulse. . . . Now the comparative Velocities of any body moved are in subduplicate proportion to the aggregates or sums of the powers by which it is moved. Therefore the Velocities of the whole spaces returned are always in the same proportions with those spaces, they being both subduplicate to the powers, and consequently all the times shall be equal.

This conclusion was of course the crux of the matter, the supposed demonstration of the isochronous quality of vibrating springs on which their suitability to control the motion of watches depended. At the very heart of the demonstration lay another conceptual confusion; Hooke drew the equality of times from the proportionality of spaces to the "velocities of whole spaces returned" in motions not only nonuniform but even nonuniformly accelerated.

Hooke was aware of course that the velocity was not uniform; and despite the uncertainty it cast on his conclusion, he attempted to compute velocity and elapsed time at any point in the oscillation. The velocity at any point is proportional to the root of the aggregate powers impressed over the space traversed. If AC represents the total strain and CD the power of the spring at C, BE will be the power of the spring at any intermediate point B (see fig. 4). The area of the triangle then gives the total aggregate of powers, and the trapezoid $BCDE$ (proportional to $AC^2 - AB^2$) represents the sum of the powers

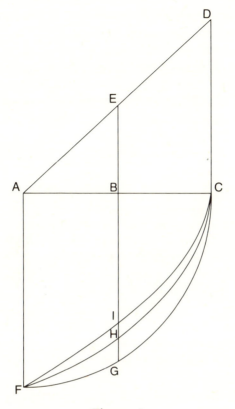

Figure 4

expended from C to B. If we draw the quadrant of a circle CGF, BG $(= AC^2 - AB^2)$ must represent the velocity at B. So far so good, but what about time? Velocity, Hooke reasoned, is in the same proportion to the root of space as the root of space is to time. The only explanation I can give for this statement is the suggestion that Hooke combined two of Galileo's conclusions for uniformly accelerated motion, $v^2 \propto s$ and $s \propto t^2$. Whatever his path, he arrived at a proportionality equivalent to the equation $s = vt$, which is valid only for uniform motion. He proceeded then to draw a parabola CHF with axis AC and vertex C. $BH \propto \sqrt{CB}$, or BH represents $\sqrt{s}$. Now draw the S-shaped curve CIF defined by the equation $BG/HB = HB/IB$, and IB gives the time elapsed in moving from C to B. Further to compound the confusion, he went on to assert that if the spring is stronger, so that its power in *C* is greater than *CD*, the circle *CGF* becomes an ellipse outside the circle and the curve *CIF* falls inside the one valid for force equal to *CD*; if the spring is weaker, *CGF* becomes an ellipse inside the circle and *CIF* falls outside.[31] Call it whatever you like, it was not scientific technology. Recall as well that at the time he composed it no scientist was in a position to correct Hooke's analysis.

I trust that no one will mistake the thrust of my argument for a denial of Hooke's contributions to technology. Those contributions are well documented. I could not effectively question them if I wanted to, and I have no desire to do so. Rather, I am examining the relation between technology and scientific investigation in the seventeenth century. Three negative examples cannot, of course, prove that the two never joined together, but surely they will lead us to exercise some caution in speaking of the existence of a scientific technology at the time, all the more so if, as I have argued, Hooke's failure in the three cases reflected the prevailing level of knowledge.

Since the dawn of the Western scientific tradition more than two thousand years before Hooke, natural philosophy and technology had been separate enterprises, each drawing upon its own sources. Should we not see in Hooke's mechanical inventions, such as the air pump, the universal joint, and the spring-controlled watch, evidence not of science at work but of a fertile imagination fed by the world of practical mechanics? By the seventeenth century, machinery of various sorts was an omnipresent reality in western Europe, to such an extent that the image of the machine could force itself on the philosophic imagination as the true representation of the physical world. What we know about Hooke's childhood indicates that machines fascinated him, as indeed they fascinated other future scientists—Newton, for example. Hooke was thoroughly at home with machinery and appears to have understood it almost instinctively. Although he pretended to call on scientific principles to support his technology, the examples I have examined strongly suggest that we should seek the foundation of much of Hooke's contribution to technology in the practical tradition of European mechanics. In his study of medieval technology, Professor White has taught us a great deal about the expanding command of machines in Europe.[32] It is significant that in his book medieval philosophers make an appearance in only one place, with reference to speculations on perpetual motion—that is, on a subject antithetical to operative technology. Hooke's example argues that the separation of the two realms, so foreign to us in the twentieth century but so common throughout Western history, had not yet ended in his age.

The seventeenth century was filled with talk of Baconian utilitarianism. Looking back at it from an era filled with the products of a technology based on the science that the seventeenth century brought into being, historians have too readily assumed that technology presented no real problems once the determination to apply science to it was born. As it happens, we have a ready empirical check on that assumption. In the seventeenth century the medical profession looked back on more than two thousand years of experience devoted to the proposition that the end of medical science is practical utility. When illness struck, however, those two thousand years proved to have delivered precious little reliable direction to the practicing physician. As we assess the application of science to technology, we ought to recall the treatment that a contemporary of Hooke, Charles II, received when he fell ill in 1685.

On Candelmas Day in the morning [Col. Thomas Fairfax, who was present, reported to the Earl of Arran] the King got up and walked above half an hour alone in his closet, and came out to be shaved; but as soon as he set down in his chair he immediately fell back in a fit; and there being by Doctor King, who had a lancet in his pocket, it pleased God to give him both courage and presence of mind to let his Majesty blood, which he did with extraordinary success, though the King's hand shook extremely. After which all the physicians and surgeons, being called in, they thought fit to apply cupping glasses and blisters to the King's head and other parts, and gave him a vomit, which worked both ways very well, and about two hours after his Majesty came a little to himself and began to speak. All that day was spent in using all means that the art of man thought proper for the King's distemper.[33]

Another witness described in greater detail the means that the art of man thought proper:

Every four hours he had a purge given him of Hyerapikra which had all the good effects wished for, he vomited four or fife tyms, but about on o'cloke this morning he had 3 or 4 large stools that so mutch relived his head and refreshed his spirits that from that time he spoke weall and as sensible as ever. Then was blisteringe plasters put to eache of his legs and thyes, to every arme, his showlders and head; about five this morninge they wear taken off and had done the opperatione extremely weall, but put his Majesty to a great deall of pain in taking off the skine, which reioiced us all extremly that he found the pain so sensibly; he slept above two hours together befor that verry soundly and quietly, and about sevin this morning he begane to talk of the way he took his diseass verry cherfully, to the unspeakable joy of all present. The Doctors then decleared he was past all hazard and this morninge about ten o'cloke thy driw some blood of him at the jugular vaens. Blessed be God he recovers hourly, and is in a verry good condition.[34]

The ministering angels of Hippocrates were not to be denied; and by diligently plying their trade, they did in the end achieve their goal. Perhaps we can use the time it took them to kill the king—four days—as a crude measure of the prevailing deficiency of medical art.

Talk of utility is one thing; the fact of utility is something else. As the example of medicine suggests, effective utility requires the accumulation of a large body of scientific knowledge. Though in Hooke's day the foundation to support such an accumulation had been laid, roughly a century had to pass before the growing store of scientific knowledge could begin to transform technology. Until that time, technology largely went its own way with its time-honored method of trial and error. For all its talk of Baconian utilitarianism, the seventeenth century was not especially rich in technological innovations.

In one particular area, however, it can be said that the seventeenth century did witness the appearance of a truly scientific technology, and Robert Hooke was among the men most prominently associated with it: I refer to the instruments that science itself used. Hooke was obsessed with the notion that the power of the senses can be extended. Observation is the foundation of scientific knowledge, but the senses themselves are weak. Their power can be enlarged by art, however, especially by mechanics, "by Means of which, not only each of them may be made more Powerful in the Discovery of the proper Objects of those several Senses; but each of them may be made a *Genus*, as it were, of new Sorts of Sense, comprised under them, of which we have yet no Notion."[35] Hooke therefore devoted attention both to the microscope and to the telescope. He consciously developed the worm adjustment to insure that the setting of a telescope could be determined with an accuracy that corresponded to the possibilities inherent in the instrument's magnification. He devised a new instrument to measure the angular distance between stars. In fact much of his catalogue of inventions was devoted to scientific instruments. Even the universal joint was conceived as part of an instrument, a clock-driven mechanism to lock a quadrant turning on a vertical axis in the azimuth plane of a star.[36] Hooke's instruments recall others and remind us that the seventeenth century was indeed productive in this one area of technology. The pendulum clock was perhaps the epitome of a technology to which we cannot deny the adjective scientific. It was not, however, a technology that answered the expectations of Baconian utilitarianism. Turned inward upon the scientific enterprise itself, it promoted the investigation of nature but did nothing to augment the comfort and convenience of life.

The identification of scientific instruments as the early focus of scientific technology prompts a new assessment of the relation of technological progress to the growth of modern science. In one prominent school of thought the two are intimately connected. Perhaps the best exposition of this point of view is found in the explicitly Marxian account of B. Hessen. "Science flourished step by step with the development and flourishing of the bourgeoisie," Hessen asserted. In order to expand its industry the bourgeoisie needed science, "which would investigate the qualities of material bodies and the forms of manifestation of the forces of nature." Although Hessen did not mention Baconian utilitarianism explicitly, he concerned himself entirely with something identical to it. The rising bourgeoisie, he argued, took science, which had hitherto been the servant of the church, into its service, "the service of developing productive forces." The rise of capitalism posed a number of technical problems in transportation, industry, and mining, and the attack on these problems "mainly determined" the work of physics. The development of productive forces, Hessen asserted, set a number of technological tasks and "made an imperative demand" for their solution. In his *Principia*, Newton, who was a typical representative of the rising bourgeoisie, presented a systematic statement of the principles on which the solutions to such problems would rest.[37]

The refutation of Hessen has been a favorite parlor sport almost from the time his article first appeared. Although I completely disagree with his argument, I do not expect to disprove his position with evidence drawn from examination of a single case. Let me only say that if Hooke's example is typical, as I believe it is, I do not see how to reconcile cases such as his with Hessen's outlook. As I indicated earlier, Hooke accepted Baconian utilitarianism as the ultimate goal of science. As an innovative technologist, however, he focused his attention, not outward on the needs of productive forces, but inward on the needs of science itself. At the conclusion of his *Animadversions on Hevelius*, a lecture devoted to the technical problems of scientific instruments, Hooke asked what purpose precision to one second of arc might have. He answered by listing nine uses, and among them, it is true, he included surveying and the determination of longitude, "a thing so highly advantageous for Trade and Navigation." Nevertheless, the list concentrated heavily on the usefulness of such precision to scientific theory. With accurate instruments science could determine the refraction of the atmosphere, measure the distances of the fixed stars, observe precisely the appulses of planets to fixed stars, and examine whether one plant influences another in its orbit.[38] If anyone was making imperative technical demands on science, it was science itself.

Let us also look at the question from the other point of view. What hardheaded businessman in the seventeenth century, rising relentlessly and cognizant of technical needs, would have thought for one moment of looking to natural philosophy for the answers to those needs? Never in the past had natural philosophy made practical utility its end. Even to voice this proposition is to impose on the seventeenth century the outlook and experience of a more recent era, despite the presence of Bacon, who was not a businessman and did not speak for them. As it appears to me, we ought rather to recognize that a unique intellectual event took place in the seventeenth century, the reconstruction of the basic categories of natural philosophy. In its insistence on the quantitative character of nature, the new science opened the door to the ultimate realization of Bacon's dream, though in a rather different way than Bacon had imagined. We today are both the benefactors and the victims of that event. The results took time to appear, however, and they came only after the reservoir of new, quantitative science had filled to a certain level. Though there may have been a few isolated exceptions, for the most part that level was not reached in the seventeenth century. To look at the technological consequences as the cause of the new science is, in my opinion, to ignore chronology and to confuse the offspring with the parent.

NOTES

1. *The Works of Francis Bacon*, ed. James Spedding, Robert Leslie, and Douglas Denon Heath, 15 vols. (Boston, 1870–1882), 8: 67–68.

2. Ibid., 40–41.

3. Ibid., 162–163.

4. Benjamin Farrington, *Francis Bacon, Philosopher of Industrial Science* (New York, 1949).

5. Richard F. Jones, *Ancients and Moderns* (St. Louis, 1936).

6. Robert K. Merton, "Science, Technology and Society in Seventeenth Century England," *Osiris* 4 (1938): 360–632. Issued as a separate volume in 1970 (New York: Harper Torchbooks).

7. Christopher Hill, *The Intellectual Origins of the English Revolution* (Oxford, 1966). Charles Webster, *The Great Instauration: Science, Medicine, and Reform, 1626–1660* (London, 1975).

8. Bacon, *Works*, 5: 398.

9. Ibid., 8: 329–346.

10. Ibid., 5: 402.

11. Ibid., 8: 481.

12. Ibid., pp. 173, 331, 333.

13. Ibid., 4: 216.

14. Ibid., 8: 169.

15. R. W. T. Gunther, *Early Science in Oxford*, 14 vols. (London and Oxford, 1930–1945), 8: 148. Vols. 6, 7, 8, 10, and 13 of Gunther are devoted to the life and works of Hooke. Vol. 8 reprints his *Cutlerian Lectures*.

16. Ibid., 8: 149.

17. Ibid., pp. 80–81.

18. Robert Hooke, *The Posthumous Works of Robert Hooke, M.D. S.R.S. Geom. Prof. Gresh. &c.*, ed. Richard Waller (London, 1705), p. 553.

19. Ibid., p. 3.

20. Gunther, *Science in Oxford*, 8: 86.

21. Robert Hooke, *Philosophical Experiments and Observations*, ed. William

Derham (London, 1726), pp. 107–108. As Derham noted, without a diagram or a model we cannot tell exactly what Hooke meant, but the intent seems clear enough.

22. *Posthumous Works*, pp. 561–562.

23. Gunther, *Science in Oxford*, 8: 163.

24. Lemma I, Bk. III, *Principia*, 1st ed., pp. 467–469.

25. Gunther, *Science in Oxford*, 8: 166–167.

26. *Posthumous Works*, pp. 565–567.

27. Thomas Birch, *The History of the Royal Society*, 4 vols. (London, 1756–1757), 2: 126. As it happens, the cycloid does fulfill this condition, though I am convinced that Hooke had no way of knowing as much.

28. Ibid., pp. 338–339.

29. Gunther, *Science in Oxford*, 8: 186–187.

30. Ibid., p. 184.

31. Ibid., pp. 349–353.

32. Lynn White, *Medieval Technology and Social Change* (Oxford, 1962).

33. Historical Manuscripts Commission, *Calendar of the Manuscripts of the Marquess of Ormonde*, 8 vols. (London, 1902–1920), 7: 316.

34. Historical Manuscripts Commission, *Report on the Manuscripts of the Duke of Baccleuch & Queensberry*, 2 vols. (London, 1899–1903), 2: 41.

35. *Philosophical Experiments*, p. 172.

36. Gunther, *Science in Oxford*, 8: 108–109, 113–143.

37. B. Hessen, "The Social and Economic Roots of Newton's *Principia*," in *Science at the Crossroads* (London: Kriga, 1931), pp. 17–20.

38. Gunther, *Science in Oxford*, 8: 111–114.

V

GUNNERY, SCIENCE, AND THE
ROYAL SOCIETY

A. Rupert Hall

More than twenty-five years ago I published a book called
Ballistics in the Seventeenth Century.[1] Today I might call it "Early
Science and the Art of Shooting." I had found a fascinating
thesis topic; it constituted not only my apprenticeship in the
history of science but also my first introduction to some of the
giants of the Scientific Revolution: Galileo, Mersenne, Huygens, Newton, Leibniz, Johann Bernoulli. I have never subse-
quently moved very far from them. The subject also drew me
into the darker mysteries of the use of guns in war.

It is a curious historical fact that while there is much historical evidence about the making of cannon, their military organization, their weights, numbers, and cost, and so on, there is
very little to show how they were actually deployed in war.
Until relatively recently, before Michael Lewis cleared it up,[2]
there was even a good deal of doubt concerning one of the
most celebrated gunslinging episodes in history, the repulse
(defeat is too strong a word) of the Spanish Armada. The
puzzle of that prolonged Elizabethan Battle of Britain turned
precisely on the characteristics and use of naval artillery; and
you may be sure that Howard of Effingham, Drake, and
others left no narrative to explain it to us. This was true of

111

nearly all battles and sieges. The technical management of guns in tactical situations was a curiously neglected branch of "baroque" military history.

There was, however, an extensive printed literature of *Buchsenmeisterei*—beginning perhaps with the German *Feuerwerkbuch* of 1420,[3] which gave rise to printed progeny branching off, on the one hand, to deal with the Brock's Benefit of what we still call fireworks, and on the other, to the great artillery handbooks of such men as Luys Collado,[4] of which Robert Norton's *The Gunner* (1628) is a late, imitative, and rather feeble descendant. These books, some of them in the form of a dialogue between "the Ancient Gunner" and "a Novice," contain a great deal of technical information on subjects ranging from the making and testing of both ordnance and gunpowder, to carriages, hoists, and horses. They instruct one in distinguishing good powder from bad, in making allowance for the decreasing thickness of metal in a piece between breech and muzzle, in determining the caliber of the shot and the charge of powder, in mastering a proper gun drill, and in handling the piece safely. Yet they say rather little about the use of the artillery in active service; they are very discreet about the range at which the various types might be employed, their placing with respect to troops of other arms, their normal rate of fire, and their effect. As for the gunners' own narratives of their experiences in actual battles or sieges, they simply do not exist.

Moreover, these early artillery manuals are highly misleading and untrustworthy in at least one important respect—and I think they share this defect with all early technical treatises. They are overelaborate and overrational. They describe the craftsman or artificer—in this case the gunner—as far too clever a fellow, weighing and measuring and calculating in a highly improbable manner.[5] Consider those many pretty mathematical books (Digges's *Pantometria* [1571], for example) in which the gunner is shown measuring the range to a castle or a ship at sea by careful triangulation or by another geometrical process involving instruments. Can one imagine that this was often done? Or that a precise measure of the range would have been very helpful in laying the gun to hit a given target? There were indeed instruments called gunner's quadrants which were supposed to be helpful in setting cannon to the right angle of elevation, but there still seems something fanciful about pictures of cannon tilted up like howitzers with their trails dug into the ground so that no recoil would be possible. Many of these laborious studies smell of the lamp; they surely contain many purely literary effects, in the manner of the navigational manuals that promised an accuracy of determination that never was attained on sea or land. And it is easy to see why. The early technological books were written not only to instruct but also to impress; and the higher their coffee-table quality, the less their didactic utility. Most were dedicated to a grandee—a monarch or successful noble commander; hence there was a patron whom the author must impress with his skill and learning. In addition, since the real gunner in the field was presumably often illiterate and very unlikely to be learning his trade from books, these handsome volumes must have been intended for the whiter hands of the gentry for whom even in the mid-seventeenth century the profession of arms was still a normal social expectation. The young gentleman had to be properly impressed by the master gunner's excellence, just as he was to be impressed in other books by the skill of the riding master, the fencing master, or the dancing master.[6] Who would wish to buy a book on a technical subject without a touch of mystery to redeem its prosaic bread and butter—the touch of mystery so blithely promised by Sir Hugh Platt in *The Jewell-House of Art and Nature* (1594) or by John Bate in his *Mysteries of Nature and Art* (1634).

The gunnery writers flattered the sophistication of their readers, who probably knew no more of triangulation, trigonometry, or even the rule of three than most real gunners; they blinded them with science. Of course no writer would ever admit openly to a bit of coney-catching, but we are told about it in satire—as in Ben Jonson's *Alchemist*, where Surley complains

That Alchemie is a pretty kind of game,
Somewhat like tricks o' the cards, to cheat a man with charming
What else are all your termes,
Whereon no one o' your writers agrees with other?
Of your *elixir*, your *lac virginis*,

Your *stone*, your *med'cine*, and your *chryososperme*,
Your *sal*, your *sulphur* and your *mercurie*,
Your *oyle of heighte*, your *tree of life*, your *bloud*,
Your *marchasite*, your *tutie*, your *magnesia*,
Your *toade*, your *crow*, your *dragon* and your *panthar*,
Your *sunne*, your *moone*, your *firmament*, your *adrop*,
Your *lato, azoch, zernich, chibrit, heautarit*,
And then your *red man* and your *white woman*,
With all your broths, your *menstrues*, and *materialls* . . .
And worlds of other strange ingredients,
Would burst a man to name?

All their talk, Subtle, the alchemist, rejoins, was used "to obscure the art" and, as his gull, Mammon, chimes in, so that "the simple idiot should not learn it, and make it vulgar."[7] But real gunnery, like other real crafts, was in fact learned by "simple idiots," ploughboys and sailor boys without benefit of letters, and we are entitled to wonder whether the alternate "literary" approach to these mysteries, which remains our only source of information, was not highly deceptive and unrepresentative of what actually happened.

The reality of the use of guns in seventeenth-century warfare is confirmed by the rich anecdotage and the pictorial evidence of a later, but still technically similar, epoch of warfare. We are all familiar with the order "Don't fire till you see the whites of their eyes," with the compliments exchanged before the battle of Minden, and with the yardarm-to-yardarm cannonades of Nelson's ships of the line. There is much to suggest the incredible resolution of combatants, little to indicate ballistic finesse.

But what of the relationship of gunnery to science in the wider sense? It has long been supposed that the various technical branches of the art of war, especially ballistics (or the science of shooting) and fortification, held the attention of men of science and mathematics, who sought for rules to define the proper practices. And indeed an interest in the flight of projectiles figures in the writings of Tartaglia, Thomas Harriot, Cavalieri, Mersenne, Galileo, Huygens, Wallis, James Gregory, Halley, Newton, Leibniz, and many more, though not necessarily in the works published in their lifetimes. In other words, the motion of projectiles, and especially the motion of projectiles in air, was studied by a considerable number of men, and by men of the highest abilities. Why? The common supposition has been—and perhaps among the less critical still is—that these men were naively eager to solve "useful" (or at least utilitarian) problems, not so much with the object of doing good (for even in the seventeenth century to increase the accuracy of shooting could hardly be regarded as clearly an act of kindly benevolence, and we have evidence that warlike inventions were then already regarded with horror) but rather in order to prove the general utility of the scientific approach to practical problems, and to demonstrate the particular powers of the individual writer.

In a classic study more than forty years ago, Robert K. Merton wrote briefly of the presence of both pure and applied aspects in ballistics research as suggesting that "to some extent at least," the interest of those who carried it out was directed to it "because of the practical utility derivable therefrom."

The effort to attain mathematical precision in artillery fire was a model for the industrial arts and a link with the current science. In any event, military needs, as well as the other technologic needs previously described, tended to direct scientific interest into certain fields.[8]

Merton's position is more subtle and more carefully qualified than this excerpt alone might indicate; for instance, he was well aware of the purely intrinsic interest of ballistic problems to mathematicians and of the force of active tradition in guiding problem choice. He was certainly far less extreme than B. Hessen, for whom science was no more than industry and commerce conducted by other means:

If we compare this basic series of themes [in seventeenth-century physical science] with the physical problems which we found when analysing the technical demands of transport, means of communication, industry and war, it becomes quite clear that these problems of physics were fundamentally determined by these demands.[9]

And indeed the military realism of seventeenth-century mechanics had been emphasized long before by William Whewell, who opined that "practical applications of the doctrine of projectiles no doubt had a share in establishing the truth of Galileo's views."[10]

In my 1952 book I argued that such views about problem choice as determined by technological need were not supported by the evidence. Existing military art was incapable of adopting a mathematical theory of projectile flight and applying it to practice, nor (so far as one can tell) did it ever attempt to do so, at least before the death of Newton. Therefore, if the learned men developed the theory out of a desire to solve useful problems, they were mistaken. It was my view, however, that they did not tackle ballistic problems for this reason but because these problems were intellectually fascinating and to some extent open to solution. I pointed out that the tradition of interest in projectile motion goes back to Aristotle; that it became a crucial issue in the medieval philosophical development of mechanics; and that it is really rather absurd to suppose that this tradition can have been immediately affected by the invention of gunpowder and the introduction of artillery. If Leonardo da Vinci was the first to bring together the post-Aristotelian tradition of academic mechanics and the brute craft of artillery, it was the mathematician Niccolo Tartaglia who first published the marriage (in 1537) and tried, without success, to make it mathematically fruitful.[11] Tartaglia's discussion was founded, on the one hand, on perfectly conventional academic ideas and, on the other, on improbable empirical facts that he unwisely supposed were certified by the experience of practical gunners. The one significant and correct idea he had about the shape of the trajectory was rejected by almost all his successors, before Galileo.

Tartaglia's recipe for practical success—to try to graft craft experience onto academic learning—failed, as it must always fail;[12] chemistry was not born from a marriage of the Aristotelian theory of elements with the experience of goldsmiths, nor was geology the product of combining miners' lore with the Book of Genesis. Galileo, for whom the theory of projectiles was an obvious extension of the most general principles of motion, relied entirely on a mathematical argument. Given that horizontal motion is uniform and vertical motion is accelerated under gravity, the parabolic trajectory necessarily follows, and from that follows an endless family of mathematical propositions which Galileo happily explored. Whether or not Galileo believed the real cannonballs or mortar bombs that combatants fired in anger actually followed parabolic trajectories is not, perhaps, very important; surely he must have had the common sense to recognize that his elaborate family of propositions was only of interest to mathematicians.

Of course, to intellectuals it was a very considerable discovery that a geometric curve was realized in physical nature, even if only under impossible conditions. Apart from the circle, only one other curve, the ellipse, was known to be so realized; and Galileo had rejected Kepler's demonstrations that it was. Galileo saw the parabola in other places too—in perfectly economical beams and in ropes suspended by their ends. And so a new issue was raised: To what extent do ordinary, actual events correspond to idealized events? Or, expressed another way: To what extent do ordinary, actual events obey idealized laws of nature? Archimedes gave one definition of a perfect fluid—How closely do real fluids approach it? Boyle's Law describes an aspect of the behavior of an ideal gas—How similar to this ideal is air? And so forth. It is one thing to declare that nature is geometrical, but (as Galileo also knew) when one comes to deal with real solids, liquids, and gases the geometry may become very perplexed.

This issue was new in the seventeenth century—or perhaps I should say it appeared new in the post-Copernican world.[13] It had not existed for the Greeks or in the medieval world—or, in fact, as long as the qualitative belonged to philosophy and the quantitative to mathematics. Light as an experience belonged to philosophy; a light ray, an imaginary abstraction, belonged to mathematics. Within such a tradition no one expected to find mathematical relationships exemplified in the real world. One recalls Simplicio's complaint: "After all, Salviati, these mathematical subtleties do very well in the abstract, but they do not work out when applied to sensible and physical matters."[14]

While the Aristotelianism of the Simplicios crumbled, the new metaphysic of mathematizing nature, sometimes called Platonism, combined with experimentally derived, quantified laws such as Snell's or Boyle's laws to reshape the problem of ideal and reality. Consider Kepler's ellipses, for example: Might not the elliptical orbit be a very close ideal approximation but the reality a complex, awkward curve barely distinguishable from the geometric ellipse? Before Newton no one could be certain, and one had to be a confirmed Platonist to assert with assurance that the geometrical ellipse is the real, as well as the ideal, path of every planet. It was for this reason that many astronomers about the middle of the seventeenth century experimented with alternatives to Kepler's version of the elliptical orbit, which Newtonian dynamics alone was later to justify and explain.

Or consider another case, that of the timekeeping pendulum. Galileo's rather informal, Platonic confidence that the ordinary circular pendulum is perfectly isochronous was proved false by Huygens in 1657; the true isochronal curve is not geometric at all but mechanical, the cycloid. What is more, Huygens proved that the center of oscillation of a mass such as a round pendulum bob is not the same as its geometric center; for all practical purposes one can measure the length of the pendulum from the center of the bob, but to be exact one must take second-order effects into account. This recognition of second-order effects as modifying, though possibly only slightly, more-or-less obvious first-order Platonic laws has been one of the ways by which science has developed.

As soon as we consider—and in due course of development carry out experiments upon—fast-moving projectiles propelled by exploding gunpowder, we hit upon an interesting example of this state of affairs. Although the first-order trajectory is clearly a parabola, the most ordinary experience suggests a whole variety of effects that must complicate this simplicity. First, as Galileo himself was well aware, a swiftly moving projectile must be slowed down by the resistance of the air beating upon it in a violent gale. Second, as tennis players know, the motion of a projectile in air may, because of inevitable asymmetries, produce changes in the trajectory. And third, the motion of the earth itself, or variations in the composition of the atmosphere, may further modify the apparent trajectory. So complex is the interplay of these various factors—all of which except the last, atmospheric variation, were recognized before the end of the seventeenth century—that there can be no completely general solution to the problem of the motion of a projectile from A to B, unless A and B are very close together.

Air-resistance, a second-order effect, attracted a great deal of attention because it was so obvious that the first-order parabola was just too naive a solution. Still, once the parabola had been proposed, it was up to the military bombardiers to make what they could of it, which was little enough, though it was to provide the official theory for mortar shooting for some time.

Rather than follow much further the contorted evolution of fluid mechanics in relation to projectiles, let me just point out that the great Scottish mathematician James Gregory, an older contemporary of Newton, solved the rather simple formulation in which the retardation is supposed to be proportional to the flight time and to act along the original direction of flight (therefore the vertical or gravitational component is supposed not to be resisted). In this case the first part of the trajectory is not much altered, but the end is squeezed in so that the range is shortened. Gregory proved that this new trajectory is itself a different parabola with its axis inclined to the horizontal to produce the asymmetric curve. Actually he was not the first to do so: Thomas Harriot did the same at the beginning of the century, but Gregory knew nothing of it as it was discovered only recently in Harriot's notes.[15]

It will be obvious that this hypothesis produces less plausible results as the parabola becomes steeper; therefore, only where the trajectory is flat can it be considered even a reasonable first approximation. In practical terms it tells us only what common sense would—that for the same range, air-resistance requires the aim to be a little higher than it would be according to Galileo's parabolic theory.

The only realistic way to analyze the problem, *pace* Gregory, was to consider the resistance of the air as acting continuously along the projectile's line of flight, directly opposing it at every point. Huygens in 1668 and Newton in 1684 succeeded in solving this mathematical problem for the case in which the resistance is supposed to vary directly as the velocity of the projectile, but both of these great mathematicians were defeated by the case in which the resistance is proportional to the *square* of the velocity, which seemed to them (rightly) more physically plausible. Newton in the *Principia* produced an approximate answer.

Thus, in a couple of generations, applied mathematics had run from the easy success of Galileo's parabola, straight into a blank wall—in the case of air-resistance only, not to mention the other complexities. No general solution of the problem of exterior ballistics is possible. Newton was fond of remarking that "Nature is always simple and consonant to herself"; and if we examine the metaphysical foundations of Newtonian science we find his confidence justified. It is in *detail* that Nature reveals the thorny problems of her complexity—whether it be in the three-body problem of gravitational force, the theory of the tides, the precession of the equinoxes, or fluid mechanics. For all his assurance of the "simplicity" of nature, there were many such problems that even Newton could not completely analyze. The hoped-for logical chain linking theory (ideal) with practice (reality) could not be completely forged. Even the near successors of Newton could not predict precisely the fall of a cannonball, any more than they could predict with perfect accuracy the exact track on the earth of the next solar eclipse, or predict to the day the return of Halley's comet. There was still an unaccountable margin of error, even if by now it was, perhaps, below one percent in most instances.

The motion of projectiles presented itself as a compelling and obvious instance of the problem of matching mathematical ideal and experimental reality. Other instances may be recalled, not least among them the related problem of fall, studied by Alexandre Koyré,[16] and the problem of longitude at sea—a valuable example because it shows so transparently the immense gulf between in-principle solutions and solutions that will give the required degree of accuracy in practice; that is, within margins of error very much less than one percent.

The problem of longitude at sea is another example of the role of practical desiderata in problem choice. No one would deny the usefulness of a workable method of finding longitude at sea; nor would anyone deny that Flamsteed's choosing to devote himself largely to the problem of lunar motion was conditioned by considerations of utility; apart from everything else, it was required of him as Astronomer-Royal at Greenwich. But suppose Flamsteed had looked less at the moon, would he have paid more attention to, say, variable stars or nebulae, or looked for new planets? Two things may be said. First, I doubt whether Flamsteed would have given as much time to the moon if the problem had not been one that interested him, and one that was capable of conferring immense prestige. The problem was clearly a scientific one. Second, it was an absorbing problem and therefore worthy of his effort. No one at this stage in the history of astronomy could find in variable stars or nebulae, or even the search for new planets, subjects likely to give an astronomer work for a lifetime. The moon was such a subject. The theory of projectiles, by contrast, offered only a limited interest. And though to some degree its problems arose from experience with projectiles and were made more dramatic by the cannon and mortars of sixteenth- and seventeenth-century warfare (firearms were high technology and remained so until Hitler built Peenamunde), insofar as projectiles won some little attention from mathematicians after Galileo, it was because projectile theory presented an intellectual challenge. I find it impossible to believe that mathematicians such as Gregory, Huygens, and Newton thought their complex curves would change the art of war.

Allow me to explore Newton's involvement a little further. Historians have demonstrated that Newton devoted very little effort to mechanics between the two plague years, 1665 and 1666, and the commencement of what was to be the *Principia* in late 1684. We know that he did some interesting and swiftly progressing work on central force during those early Cambridge years, when he was also most active in mathematics, and that he touched on some other topics in mechanics also; but nothing more happened until Robert Hooke opened a correspondence with Newton in 1679. There is one incident we should attend to, however. In 1674 Robert Anderson published a book on gunnery. Although he was apparently not without mathematical learning—his was the first gunners' manual to make use of logarithms—Anderson clung in an already naive way to the simple parabolic theory. He was quite prepared to contemplate ranges of ten miles or more (then wholly imaginary) and still suppose the projectile to be unaffected by its passage through so much air. He was criticized for this by James Gregory, who produced in rebuttal the inclined parabola theory. Gregory's London friend John Collins, the correspondent of so many mathematicians at home and abroad, sought to secure support for Gregory's poor opinion of Anderson's book by soliciting the opinions of mathematicians such as John Wallis, Savilian Professor at Oxford, and Isaac Newton, Lucasian Professor at Cambridge. It seems now rather like taking several steam hammers to a small nut. Both Wallis and Newton condemned Anderson. Fortunately we have Newton's letter of reply, pointing out precisely the probable error on which Gregory had insisted. Remarking that according to the parabolic theory the projectile must actually accelerate in the downward half of the trajectory, despite air-resistance, he gave it as his view that such an acceleration was very unlikely and proposed (arbitrarily) an alternative construction representing the projectile as continually slowing down throughout its flight, which he thought it must do in practice. However, he told Collins not to mention his name if he passed this opinion on to Anderson, "because I have no mind to concern myself further about it."[17]

These words proved a poor prophecy, however, for when his thoughts were returned to mechanics by Halley's visit to Cambridge in August 1684—and the famous question "What curve is described by a planet under the action of an inverse-square central force?"—Newton spontaneously tackled the problem of resisted motions in the *Propositions on motion* he drew up that autumn.[18] In the *Propositions* Newton introduces the matter as an aspect of a quasi-Aristotelian distinction between celestial and terrestrial motions. Only in the heavens, he points out, can motion accord with the abstract mathematical propositions outlined in the first part of his paper, "For I think that the resistance of pure aether is either non-existent or extremely small." On earth all real motions take place in fluids, and their representation by geometry becomes more complex. Newton then goes on to show how this may be done by hyperbolic constructions on the assumption that the resistance is directly proportional to the velocity (these are equivalent to the results Huygens had obtained in 1669, using the related logarithmic curve). And, he continues, the same method of analysis may be applied to movement in water.

It seems obvious that in these latter propositions, the foundation of the second book of the *Principia*, Newton went far beyond Halley's original request for a dynamic derivation of the planetary orbit. In the *Propositions*, and far more elaborately in the *Principia*, Newton embarked on a general account of rational mechanics appropriate to both earth and heaven, air and ether. Among the "cases" of motion in air that he chose to examine were the oscillation of pendulums, free-fall, and projectiles, cases chosen because they were so well known as to be notorious, and because they offered possibilities of comparing theoretically computed and experimentally measured results. There is little or no practical value in discovering how the motion of a clock pendulum is affected by air-resistance or how much more slowly a stone falls in water than in a vacuum. With Newton, then, the problem of the trajectory of a projectile entered the highly mathematical realm of rational mechanics, where it rested for a very long time. I shall not disturb it further.

If no one was about to put Newton's mechanics to practical test by firing ordnance, it was far different with Galileo's parabolic theory. Here the issue was to find out how far, if at all, theory was faulted by experience. Some of the more practical men in the Royal Society seem to have been upset by James Gregory's assertion, against Anderson, that the parabola theory could not hold in actual shooting, while Anderson himself seems to have accused the king's gunners of the Tower of ignorance, or some other negligence now obscure. At any rate, in September 1674 Collins, in London, was able (with an undertone of amusement) to report to his friend Gregory in Edinburgh that the book had "caused much bustle which hath caused the Master of the Ordnance, Sir Thomas Chichele, Sir Jonas Moore, The Lord Brouncker and the gunners [of the Tower] to go many times a shooting with mortar pieces at Blackheath." Another person involved was Robert Hooke, who had bought Anderson's book on 20 May and who went with those just mentioned to Blackheath on 11 September for the shooting, on which date he notes, "Dined on Lord Brouncker treat and coach." On 17 September they went again: "Tried experiment of bullet with good [results] found it very near a parabola."[19] How the tests were made is not stated; it was presumably done by keeping the powder and shot as constant as possible and seeing if the ranges at different angles of elevation came out as predicted by the tables of parabolas.

Sir Jonas Moore, a latecomer to the Royal Society (he was to be elected only in December of the same year, 1674), was surveyor-general of the ordnance and therefore profession-ally interested in the Blackheath trials; similarly Lord Brouncker, president of the Royal Society, was a member of the Navy Board, an associate of Samuel Pepys in fact, and was therefore also involved in the defense of the realm. Moreover, Brouncker had long before made some experiments on explo-sive recoil which had been published in the Royal Society's *History*.[20] We may presume that in the case of Brouncker a mathematical interest in ballistics—he was a mathematician of some note—combined with professional position. The pres-ence of Hooke at these Blackheath trials and also that of Henry Oldenburg, secretary of the Royal Society, is a little strange unless we presume that Brouncker wished to lend scientific color to the trials by bringing along two of the soci-ety's officers. Equally likely is that he thought it would be an agreeable day's excursion for his two friends, for the society was in recess, not meeting again until 12 November, and no report of these affairs was ever made to it.

The trials at Blackheath were by no means unprecedented in the Society's history. Nearly ten years previously an entry in the Society's Journal Book indicates that a distinguished of-ficer, Colonel George Legge (1648–1691)—in later years to be lieutenant general of the ordnance and commander-in-chief of the Navy as Baron Dartmouth—was to be thanked by Sir Robert Moray, vice-president, for having obtained from the king a gun for the society to make experiments with; and Robert Hooke, the society's curator of experiments, was in-structed, rather late in the day one might suppose, to draw up a series of experiments for the improving of artillery. Why the use of the gun was sought, and whether the experiments were ever attempted, we do not know, since nothing more on the subject appears in the Journal Book. Nor do we know whether the list of artillery experiments proposed by Sir Robert Moray and published in the *Philosophical Transactions* in 1667 had any connection with these events.[21] Moray proposed to investigate three principal "practical" points: (1) the point-blank range; (2) the optimum quantity of powder in the charge for full range; and (3) the length and bore for maximum range. Internal evidence makes it clear that the writer had little or no real experience of artillery and its use.

Though at this time the society's attention was instantaneously diverted from cannon to onions and the vegetation of plants, it was not always so capricious. At its very beginnings, during 1660 and 1661, it had repeatedly urged Lord Brouncker to make those experiments on the recoil of firearms previously mentioned, until at last they were carried out in the courtyard of Gresham College.[22] Brouncker seemed intent to prove that if recoil caused the barrel to deviate from its "laid" line, as in those days was commonly the case, the shot would be forced out of its true course by the pressure of the barrel on one side or the other. It was perhaps an obvious point, but he proved it rather neatly by the experiments—as also that the velocity of the shot could alter the manner of its being diverted by the recoil.

A few years later, there was a sustained period of interest in the propellant of firearms, ordinary black powder, and its comparison with other explosives, especially *aurum fulminans*, fulminating gold. The problem immediately raised was, How can one explosive be experimentally compared with another? especially since an "explosion" was assumed to be by no means a simple, repeatable phenomenon. Part of the effort, therefore, was expended on improving "testers" of gunpowder, devices in which a few grains of explosive might be fired to blow up a heavy hinged lid or a plunger to a measured height. Hooke devised an "improved" form that did not fully justify itself since it all too frequently shattered. It did emerge that a black powder of the ordinary kind but made with extraordinary care and refinement, such as that Prince Rupert provided the society with a specimen of, could be as much as ten times more powerful than commercially made powder.[23]

The strange behavior of explosives when shut up in solid steel was also explored, though for what purpose is far from clear unless it was in order to direct an irresistible force against an unbreakable object.[24] It is not without interest that in connection with these experiments Prince Rupert sent in a report on the efficacy of gunpowder in blasting rock—an early reference to explosive quarrying—and Hooke alluded more than once to his ambition to use gunpowder to bend a spring, which, as he said, would enable a man to command force without limit.[25] These trials did not succeed; but we may perhaps connect them with Huygens's later interest in the use of explosive force to do mechanical work, brought back to England by Denys Papin. Of course Huygens invoked the indirect principle by making atmospheric pressure do the work, which leads to the steam engine tradition.

The interest in gunpowder and explosions was certainly related to the contemporary preoccupation with nitre or salt-petre, which makes up about three-quarters of black powder. Was not nitre, or rather a nitrous component in the air, an essential agent in combustion and respiration and, very likely, in the nutrition of plants? So, at least, Hooke and others were prepared to assert and demonstrate by experiment. To follow this clue would lead us far afield, but if we pursued it we would get a full picture of why on 15 February 1665 Hooke made an experiment to show that gunpowder would burn in vacuo, as Robert Boyle had previously asserted it would, and why a fortnight later other trials were made on the addition of sulphur—the other "fatty" principle of combustion—to hot nitre.[26] The fact is that this villainous fellow gunpowder was thought to partake of the magical quality of nitre, which at this time was cast in the role of the principle of life.[27]

Discussion of this question recurred time and again in the Royal Society's debates,[28] as did that other hardy perennial, the effect of air-resistance on the movement of projectiles. On the whole, the society tended to support Galileo's position. It showed no inclination to doubt his assumption or axiom that a curved motion can be considered as compounded from two rectilinear elements, or (to put it more simply) that the horizontal, uniform component in the motion of a projectile is undisturbed by the vertical, accelerated component. This axiom had been rather crudely verified by the Accademia del Cimento in Florence and adopted by Borelli in *De vi percussionis* (1667).[29] Nevertheless, its strong a priori character had been challenged by Honoré Fabri in *Dialogi Physici* (1669), and the challenge was reported to the society on 24 November 1670. Though the general presupposition was in favor of Galileo and Borelli, the Royal Society thought that an experimental trial should be made, though it did not contemplate field trials with actual artillery such as the Florentines undertook.[30] On 8 December Hooke was "ordered to prepare for an experiment to be made at the next meeting in the assembly room, by having two balls, and projecting the one horizontally from the window over the door, and letting the other fall down perpendicularly from the same height." This was actually done on 26 January 1671 and twice repeated, but there was always doubt about whether the two balls hit the floor simultaneously, and though most of those present thought they did, the question was never decisively settled. As with so many other matters, the general opinion prevailed and became customary because it seemed reasonable rather than because exact experiment falsified the opposite belief.[31]

The effect of air-resistance on the trajectory that, on the basis of this first approximation, was deemed parabolical, was a harder nut to crack, not only in the more refined mathematical analysis already reviewed but also in the experimental or commonsense approach. Hooke raised the question as early as March 1663, in a rather confused series of proposals that came to nothing at that time. When he and Walter Charleton were ordered in the following year to report on the measured velocity of bullets, they came back with the figure 720 feet per second, surely too low.[32] Nothing more happened for several years, until in 1669 at the time when the society was investigating the laws of impact and the "force" of moving bodies, it was suddenly proposed as a possible experiment, "To show the proportion of the resistance of the air to bodies moved through it."[33] It can hardly be a coincidence that, in Paris in the previous year, Christiaan Huygens had begun to explore the same problem, and a link is furnished by the regular correspondence between him and Sir Robert Moray in London.[34] The London experiments were made with a flat pendulum: the larger the arc it swung in, the more it seemed to suffer resistance, "and the impediment to motion decreased in a greater proportion than the decrease of the velocity," though the exact proportion was not discovered, and repetition of the experiments threw no clearer light on the phenomenon.[35]

There followed another interval in which nothing happened. Then, in July 1675, Hooke once again showed an experiment on air-resistance, which may, however (the minute is laconic), have related more to his notions about flying and the action of wings than to projectiles;[36] after this, three more years elapse before the topic crops up again. Anderson's book and the problems it raised were not touched on by the society—perhaps because the "summer" vacation was prolonged that year until 19 November—although John Wallis had written at some length from Oxford of his reasoned rejection of the simple parabolic hypothesis (joining with Newton in this), in letters to John Collins and Henry Oldenburg.[37]

There could be no question but the theoreticians of mechanics took the matter of air-resistance seriously, and now Robert Hooke himself at last spoke out openly to the same effect. Challenged in his belief that a marine depth-sounder, in its descent to the bottom and ascent to the surface, would travel distances (depths) proportional to the elapsed time, he retorted by explaining the notion of "terminal velocity," illustrated by the guinea and feather experiment. "All mediums whatsoever [he said] had some resistance to the motion of bodies through them, and even those which had least, had yet a very considerable opposition to a motion that was proportionably accelerated. Hence it was how birds were able to sustain themselves in the air. . . ." Hooke's friend Sir Jonas Moore (now risen under the new regime to be vice-president of the Society), who had once gone to Blackheath to test the parabolic theory, now recalled that in shooting mortar bombs the greatest range was at somewhat less than forty-five degrees, since the shots at high elevations were more deviated from the exact parabolic line than those at lower elevations; and he expatiated on the atmospheric effect on projectiles.[38]

If the Society's corporate contribution to the study of the flight of projectiles seems uncertain, repetitive, and inconsistant, especially as regards the experimental mode of inquiry, it must be remarked that this is no more than typical of any topic one tries to follow through the Society's Journal Book. Rarely was a topic examined in an exhaustive, or even a systematic, manner; rarely was even a tentative conclusion reported, except in the form of passive recognition of the results of work carried out elsewhere by an active Fellow such as Boyle or

Grew. So often an investigation was begun and dropped after a couple of meetings; or the order was given to "let this experiment be shown" and it was not done; or the apparatus worked indecisively or not at all and the whole matter was abandoned. Out of many instances, the example of the reflecting telescope is notorious, for it is obvious that common sense, physical labor, and enthusiasm could have created a successful large-scale reflector in the 1670s just as certainly as in the 1730s.

The Royal Society was, in fact, a very inefficient agent of experimental research. Baconianism was easier to preach than to practice and led too easily into whimsy and incoherence. In the painfully slow evolution of experimental science we should certainly note, however, that the more serious of the Fellows were conscious of the Society's shortcomings. At the end of 1666 Sir Robert Moray was already urging the Society to take stock, to consider "how the experiments at the public meetings of the Society might be best carried on; whether by a continued series of experiments, taking in collateral ones, as they were offered, or by going on in that promiscuous way; which had hitherto obtained." Just over two years later, no improvement in method having occurred meanwhile, the secretary, Henry Oldenburg, renewed the question, moving that the Council "would think upon an effectual way of carrying on the business of experiments at the meetings of the Society" and that committees be appointed "proper for directing of experiments"—that is, for planning a series of them in advance. Oldenburg's motion was carried and the president said he would see to it, but experimentation remained as promiscuous as before.[39]

In the 1670s, as the Society's financial position worsened (for although the list of Fellows was long, few paid their subscriptions regularly) and the laughter of its critics grew louder (it was notorious that "very many things were begun at the Society, but very few of them prosecuted"), it seemed more than ever necessary to systematize its experiments and discourses.[40] In the ponderous words of Sir William Petty, "The Society has been censured (though without much cause) for spending too much time in matters not directly tending to profit and palpable advantages (as the weighing of air and the like). . . ." If we may judge from Petty's proposals for reform in 1673, even such apparatus as the Society had managed to assemble for making "the experiments of motion, optical, magnetical, electrical, mercurial etc. And such instruments as had formerly been used by the Society" were disorganized, scattered, and broken.[41] Shortly thereafter, with Petty apparently still leading the party of reform and action, a purge of useless, nonsubscribing Fellows was instituted and a committee set up to prepare "a list of considerable experiments to be tried before the Society" and to collect the necessary apparatus.

The outcome, though not wholly negligible, was clearly far from satisfactory. Several months later Petty again lamented the Society's "want of good experimental entertainment at their meetings," which was a principle cause of its sluggish condition. He now proposed that a team of Stakhanovites volunteer to submit an experimental discourse or considerable experiment, in turn, week by week, under pain of a fine for nonperformance. This was done for a time, and with the appointment of a second curator, Nehemiah Grew, things looked up—only to lapse once more into inanity after the Hookian Thermidor following the death of Oldenburg in 1677, when Brouncker and the old guard were ousted. Years later even such young Turks as Edmond Halley or Denys Papin were neither insistent nor successful in causing the Society to investigate in accordance with a rational research program.[42]

There was no institution, no model, no method, for systematic experimental science. Only in observatories was work regularly pursued according to some strategy, clear night after clear night, and even then the program was likely to be interrupted or switched by some chance event such as the appearance of a comet or a big sunspot. Only individuals—Boyle, Newton, Grew—were able, and inspired, to follow a line of investigation with drive and persistence, yet rarely for longer than a few weeks on end. How often does Boyle report the loss of some experiment because of the untoward arrival of visitors, just as Newton himself left his optical experiments incomplete. Set in this context, the launching of the elaborate and sustained experimental investigations that alone would yield permanent results, in the study of fluid motion generally and of projectiles in particular, may be seen as quite hopelessly impossible. What the Royal Society actually achieved was no more than the feeblest gesture of faith in empiricism.

After all was said and done, despite Bacon and *Nullius in Verba*, the Royal Society was as apt as anyone else to believe what it wanted to believe, as the slightest inspection of Birch's *History* will demonstrate. The antischolastic skeptic could too easily become the credulous mechanist, for the mechanical philosophy could as well render up false or misleading answers as could Aristotle's philosophy. And in the case of projectiles, the a priori mechanist could be doubly deceived. He was deceived in the first place if he accepted Galileo's parabolic theory as more than an approximation, imagining air-resistance to be negligible; but he was deceived again, more subtly, by the logically consequent supposition that resistance obeys some simple law—that it must be directly proportional to the velocity, or to the square of the velocity. How seductive the square-law was may be seen from Sir William Petty's treatise on it. On wholly a priori grounds Petty saw the duplicate proportion as everywhere, a fundamental law of nature having far-reaching technological implications. For example, he stated that in ships the speed attained was as the square root of the driving power, which he equated with the sail area; which is to say that the resistance increases as the square of the velocity. By similar reasoning he put the speed of a bullet as the square root of the amount of powder fired. Huygens, Newton, and many others adopted the square-law as virtually self-evident, though Newton admitted it only as an approximation. Why experiment when nature offered so ready and satisfying an answer?

Thus the matter was handed back to the mathematicians, with neither experiment nor experience pronouncing an effective word, and the problem became one of defining the trajectory curve, the reshaped parabola, in terms of a sharply formulated a priori law of resistance. In this form (as we have already seen) the problem could be solved when the resistance was made to vary directly as the velocity; it was insoluble, however—in exact and general terms—for the case of proportionality to the velocity squared. Newton examined an approximation to the unattainable exact solution, but—and this should be emphasized—the mathematical solution (or nonsolution) to the mathematician's problem was only of interest to mathematicians. The a priori character of the assumptions on which the problem of the trajectory curve was based entailed this consequence: because experiment, not to say experience, had been thrust aside, ignored, or found unnecessary in defining the problem, the problem's solution could not relate to experiment or experience, save by coincidence. Moreover, the physically "better defined" trajectory was not only nonexperimental, it was useless too.

This is an example of applied mathematics ascending into a stratosphere of sophistication which is far beyond the sight of ordinary mortals, and from which it is unable to return with any workable proposals. We have to remember that even relatively obvious questions in mechanics, such as that of the height of a jet of water spouting upward in relation to that of its pressure-head, could contain great complexities of theoretical analysis.[43] In this case (the case of *Principia*, book two, proposition 37 of the first edition) Newton was saved from serious errors of reasoning and fact by his editor, Roger Cotes.[44] In the case of projectile motion Newton was less fortunate, and an initial unfortunate slip (there was no error of reasoning) led him into difficulty and embarrassment.[45] No wonder, then, that these problems came to be of purely technical, mathematical interest and of no possible practical import.

Consider also Christiaan Huygens's mathematically elegant but practically clumsy solution to the problem of timekeeping—making the pendulum swing between cycloidal arcs so that it too would trace a cycloid. The mathematical elegance of this design is redundant, since it is enough to reduce the pendulum's swing to a small arc of a few degrees, as was done a few years later with the anchor escapement in clocks. In the small arc, the difference between cycloid and circle is both theoretically and technically negligible.

In ballistics the effect of air-resistance can never be ignored without a corresponding loss of precision. As the more complex rational approach to the problem led only to sterility, however, there was only one alternative to giving up the whole problem as insoluble: to adhere to the original rational solution provided by Galileo in the parabolic projectory. With this simplification accepted, and air-resistance therefore ignored, all that remained for the mathematical ballistician was a simple exercise in geometry—resulting in paper constructions or instrumental procedures developed for the supposed benefit of the practical artilleryman. Since at this time weapons were either guns fired more-or-less at point-blank range or mortars shooting off bombs at high elevations but with little pretense to close accuracy in grouping, parabolic tables were probably as practically useful as any that could have been generated in accordance with more sophisticated hypotheses. The mortar bombardier could, if he took the trouble, at least derive comparative elevation-range correlations from them.

It is well known that the most impressive French study of exterior ballistics at this time, François Blondel's *L'Art de Jetter les Bombes* (written in 1675 but not published until 1683 for national security reasons), was couched wholly in terms of the parabola. It nevertheless contained contributions from some of the best scientific minds working in Paris—Jean-Dominique Cassini, Ole Roemer, Philippe de la Hire, and C. F. Milliet de Challes. Furthermore, Blondel's tone was eminently "practical"; though he was aware of the parabola's status as an ideal approximation to the true trajectory in air, the parabola was, he reported, confirmed by ocular tests. It was therefore worthwhile to handle expeditiously the problem of finding elevations when shooting over sloping as well as horizontal ground. This was the chief business of his book, which the mathematicians dealt with in a variety of ways.

Edmond Halley, following Blondel's track, considered this same problem in two papers in the *Philosophical Transactions* (of 1686 and 1695). He was the first to point out that the greatest range is always obtained when the angle of fire bisects the angle between the ground and the vertical;[46] but his main mathematical interest was in the problem of shooting over rising and falling ground, a problem which Halley mistakenly said had not been treated by Torricelli and whose first solution had been given by Anderson. Actually Torricelli *had* solved this problem and with a construction more elegant and simple than Halley's.[47] Like the French mathematicians, Halley argued that "the opposition of the air . . . in large and ponderous shot . . . is found by experience but very small, and may safely be neglected." Thus the parabolic tables could be safely relied on, he thought, "as if this impediment were absolutely removed." He supported this view by a rather vague allusion to some trials made with a small mortar. Halley's early nineteenth-century editors rebuked him for giving this assurance, pointing out that, were the ball not impeded by the air, the range of cannon would be ten or twenty times as great as it actually is.[48] The criticism is perhaps not entirely apropos since Halley himself was well aware that, as applied to "great guns" with a velocity well in excess of 1,000 feet per second, "the theory of Galileo allowing no opposition to the ball from the air is insufficient in great distances."[49]

We must, moreover, credit Halley with the common sense to perceive that tables were not a great desideratum of the field gunner if computed with "an exactness and curiosity beyond what the gross practice of our present cannoneers seems to require, who lose all the geometrical accuracy of their art from the unfitness of the bore to the ball and the uncertain reverse [recoil] of the gun, which is indeed very hard to overcome."[50] This last comment was written in 1701, after two generations of familiarity with the parabolic trajectory, and indicates that the situation then was just as it had been in the beginning: geometrized shooting was an ideal unattainable in practice.

What are we to make of all this curious miscellaneous activity? We see first of all the natural eagerness of the Royal Society's Fellows and of their fellow academicians in Paris—and indeed of numerous private mathematicians—to demonstrate that their science could cope with utilitarian problems. I need hardly recall the parallel work in the geometrization of fortification by the engineers of whom Vauban is the best known.[51] Scientific research had been commended to the state not only because science brings intellectual enlightenment but also because it brings fruit, and some scientists, at least, thought it incumbent on them to justify this promise. As so often happens, propaganda had pronounced a self-fulfilling prophecy.

In addition, we can observe, I think, a consistent distinction between the "practical" men, such as Brouncker, Moray, Petty, and Moore—though Charles II's shipwrights would hardly have called Petty practical—some of whom actually carried out artillery experiments, and the mathematical idealists, such as Gregory, Huygens, and Newton, all of whom regarded the parabolic theory as in principle too simple in its physical assumptions. Halley's work comes between these two positions. The latter group I think we may discharge of any concern for actual shooting; in an age of experiments a man who produces a paper ballistic theory of great mathematical complexity can scarcely be regarded as a utilitarian. For the mathematicians, surely, the interest was in puzzle solving. The former group were more intent on practical accomplishment—this stands out in Moray's 1674 paper—and though we may, with hindsight, doubt whether their efforts could in any circumstances have proved fruitful (for the reasons noted by Halley), the explanation of their failure to push on steadily with the accumulation of performance data lies more in the unsystematic character of experimental science at this time, and its inadequate control of material resources, than in a lack of interest in such an approach. Social reasons, one may surmise, obstructed the social pursuit of science; the mathematical exploration of the problems of gunnery was not similarly limited and was pressed much further forward.

It is curious how the parabolic theory, at its inception so highly scientific and mathematical, rapidly became the standard resource of those who would geometrize shooting. This translation from the scientific frontier to the commonplace world through a process of simplification and codification represents in miniature what was commonly to take place later in the application of science to technical problems and in the education of technical operators. The parabolic theory provided very simple, schematic solutions to all problems of gunnery; it was difficult, if not impossible, to prove that it was seriously in error for all practical situations in the field; and accordingly it was widely, and for long, publicized by the would-be instructors of cannoneers. It is still difficult to determine whether it really influenced practice, however, and I must leave that problem, if it can be solved, to others.

As for the relative significance of these investigations and aspirations within the whole compass of seventeenth-century science, opinions will probably differ forever. I think it is useful to make the sort of distinction between individuals that I have already indicated—and we may at least be sure that in the whole scientific work of a Huygens or Newton the theory of projectiles was a tiny element. I am inclined to think that if we view the work of the Royal Society as a whole we shall not come to a very different conclusion, despite the not uncommon opinion (which Robert Merton is held to have proved) that many of the early interests of the Royal Society revolved around the problems of mining, navigation, and war. In fact, Merton only made a claim that 7.2 percent of his sampling of Birch's *History* related directly to military technology—in which reckoning he may have included all references to air-resistance, the laws of impact, nitre, and so forth.[52] One can only suppose that those who suggest that those three technical issues were constantly, in a technical or utilitarian sense, the object of the Society's attention have spent little time looking through Birch's volumes.

I shall not hesitate to agree with Merton's conclusion that some influence on problem choice was exerted by practical questions such as gunnery; with whose problems a few of the Fellows were plainly well acquainted, though I think Merton's quantitative measure of this influence was high. Once the problem has been chosen, however, may not the historian also ask what contribution the investigation of it may be seen to make to either technique or knowledge—what, in Bacon's words, does it yield of "fruit or light"? In the case of gunnery, the yield seems slight as to knowledge—it contributed to a minor branch of rational fluid mechanics—and nil as to technique. One cannot believe that the Fellows of the early Royal Society—from Ashmole to Wren—would sustain their reputations by such sporadic, slight, and ineffective investigations as those I have recounted here.

The fact is that the investigations that bore fruit were in the quite different areas of astronomy, chemistry, optics, pneumatics, and so forth. And it was necessarily so. The scientific mastery of the complex problems of technology still lay a century or more in the future because it depended on two things: the elucidation of clear scientific models (in chemistry, for example) and the appearance of men, such as Watt and Smeaton (or Böttger and Macquer), who would successfully investigate technical problems in a scientific manner, as was almost never done in the seventeenth century.

The notion that technology is simply "applied" or borrowed science has been rightly criticized by historians of technology. We now believe that there is more to the solution of technical, and indeed, scientific, problems than the awakening of interest in their solution. The history (or sociology) of problem choice is a significant study, but it is well to remember that the selection of a problem for study (and those words themselves beg a lot of questions) is only a necessary, not a sufficient, condition for the provision of a usable solution.

NOTES

1. A. R. Hall, *Ballistics in the Seventeenth Century* (Cambridge; At the University Press, 1952; reprinted, New York, 1969). The historical literature has since been enriched by a number of works treating of the history of artillery from the economic, strategic, and political point of view, such as Carlo M. Cipolla, *Guns and Sails in the Early Phase of European Expansion, 1400–1700* (London, Collins, 1965) and John Francis Guilmartin, *Gunpowder and Galleys* (Cambridge, University Press, 1974). Guilmartin especially has provided excellent technical material on sixteenth-century cannon and their use, confirming my contention that long ranges and hence ballistic tables were irrelevant to cannon fire in warfare, save in exceptional circumstances. I should note also H. J. Webb, *Elizabethan Military Science: The Books and the Practice* (Madison, University of Wisconsin Press, 1965).

2. Michael Lewis, *Armada Guns: A Comparative Study of English and Spanish Armaments* (London, 1961).

3. W. Hassenstein, ed., *Das Feuerwerkbuch von 1420: 600 Jahre deutsche Pulverwaffen und Buchsenmeisterei* (Munich: Verlag der Deutschen Technik GMBH, 1941).

4. Luys Collado, *Pratica Manuale di Arteglieria* (Venice, 1586).

5. I do not mean that it is improbable that cannonballs and charges of powder were not measured out with some care; this prudence would demand. I refer only to various hypothetical procedures ancillary to, or involved in, laying the gun.

6. Guilmartin has shown that the range tables often given by the early writers on artillery (some examples of which were printed in the appendix of my book *Ballistics*) are quite impossible and would sometimes require quite unattainable muzzle velocities; all of which strengthens my point.

7. *The Alchemist*, Act II, Scene III, 11.184–202.

8. Robert K. Merton, *Science, Technology and Society in Seventeenth Century England* (*Osiris* 1938; reprint, New York: Harper Torchbooks, 1970), p. 191.

9. B. Hessen, "The Social and Economic Roots of Newton's *Principia*," in *Science at the Cross Roads* (London, 1931; reprint, London: Cass, 1971) p. 166.

10. W. Whewell, *History of the Inductive Sciences* (1837), quoted in Merton, *Science, Technology and Society*, p. 187.

11. Niccolo Tartaglia, *Nuova Scienza* (Venice, 1537); *Quesiti et inventioni diverse* (Venice, 1546).

12. Alexandre Koyré, "La Dynamique de Niccolo Tartaglia," in *La Science au Seizième Siècle*, Colloque de Royaumont 1957 (Paris: Hermann, 1960), 93−113.

13. The *locus classicus* is Pierre Duhem, *To Save the Phenomena: An Essay on the Idea of Physical Theory from Plato to Galileo* [1908] (English translation, Chicago: University of Chicago Press, 1969). See also Edward Grant, "Late Medieval Thought, Copernicus, and the Scientific Revolution," *Journal of the History of Ideas* 23 (1962): 197−220.

14. Galileo Galilei, *Dialogue Concerning the Two Chief World Systems*, trans. Stillman Drake (Berkeley and Los Angeles: University of California Press, 1953), p. 203.

15. Galileo Galilei, *Two New Sciences*, trans. Stillman Drake (Madison: University of Wisconsin Press, 1974), p. 223−229: "I admit that the conclusions demonstrated in the abstract are altered in the concrete, and are so falsified that horizontal motion is not equable, nor does natural acceleration occur in the ratio assumed; nor is the line of the trajectory parabolic and so on." Gregory's work is in "Tentamina quaedam geometrica de motu penduli et projectorum," annexed to William Saunders, *Great and New Art of Weighing Vanity* (Glasgow, 1672); Harriot's is in British Museum Add. 6789, f. 69r. (See D. T. Whiteside, *Mathematical Papers of Isaac Newton* [Cambridge: At the University Press, 1974], 6:7, n. 16 and 17; I am also indebted to Dr. Whiteside for a private communication.)

16. Alexandre Koyré, "A documentary History of the Problem of Fall from Kepler to Newton," *Transactions of the American Philosophical Society* 45 (1955): 329−395; and idem, "Galileo's *De motu gravium*" in *Metaphysics and Measurement* (Cambridge, Mass.: Harvard University Press, 1968), pp. 82−88.

17. Newton, *Correspondence*, 1: 309 (20 June 1674); see also H. W. Turnbull, *Gregory*, p. 282 ff. (note 19 below).

18. See A. Rupert Hall and Marie Boas Hall, *Unpublished Scientific Papers of Isaac Newton* (Cambridge: At the University Press, 1962), pp. 287−292; D. T. Whiteside, *Mathematical Papers of Isaac Newton* (Cambridge: At the University Press, 1974), 6, 65−75.

19. H. W. Turnbull, *James Gregory Tercentenary Memorial Volume* (London: G. Bell & Sons, 1939), pp. 282, 286; H. W. Robinson and W. Adams, eds. *The Diary of Robert Hooke, 1672−1680* (London: Taylor & Francis, 1935), pp. 103, 121.

20. Thomas Birch, *The History of the Royal Society* (London, 1756), 1:20, 33; the experiments were made on 7 April 1661 and presented to the society on 10 July; see also Thomas Sprat, *The History of the Royal Society* (London, 1667), p. 237 and Christiaan Huygens, *Oeuvres Complètes* (The Hague, 1890), 3:323.

21. Birch, *History of the Royal Society*, 2:24 (15 March 1665); *Phil. Trans.* II, no. 26 (3 June 1667): 473–477. It is not impossible that Hooke had some hand in this list.

22. Ibid., 1:20, 33.

23. Birch, *History of the Royal Society*, 1: 281–5, 295, 299, 302–303, 335 (22 July–25 November 1663).

24. Ibid., 1: pp. 425–465 passim (May–September 1664).

25. Ibid. 1:335, 2: 446 ff. (January–February 1667).

26. Ibid. 2:15, 19.

27. Shakespeare, *Henry IV*, Pt. I, I, 3, 11.60–63; D. McKie, "Fire and the *Flamma Vitalis*," in E. A. Underwood, ed., *Science, Medicine and History* (Oxford: Clarendon Press, 1953), I: 469–488; Henry Guerlac, "John Mayow and the Aerial Nitre," *Actes du 7 Congrès d'Histoire des Sciences* (Jerusalem, 1953), pp. 332–349; idem, "The Poet's Nitre," *Isis* 45 (1954): 243–255 (both reprinted in his *Essays and Papers in the History of Science* (Baltimore: Johns Hopkins University press, 1977).

28. E.g., Birch, *History*, 2:468 ff. (13 March 1679 and subsequent meetings) leading into a discussion of Roger Bacon's role in the invention of gunpowder and its use in artillery.

29. Galileo, *Dialogue* (note 14), p. 155; Richard Waller, *Essayes of Natural Experiments* (London, 1684), p. 143; W. E. K Middleton, *The Experimenters* (Baltimore: Johns Hopkins University Press, 1971), pp. 240–243; G. A. Borelli, *De vi percussionis* (Bologna, 1667), p. 169.

30. Birch, *History*, 2:454; Wallis to Oldenburg, 22 November 1670, in A. Rupert Hall and Marie Boas Hall, eds., *The Correspondence of Henry Oldenburg* (Madison: University of Wisconsin Press, 1970), 7:283–284.

31. Birch, *History*, 2:461, 465, 467.

32. Ibid., 1:205 (4 March 1663), 465 (7 September 1664), 474 (5 October 1664).

33. Ibid., 2:339 (14 January 1669).

34. Hall, *Ballistics*, 111–113.

35. Birch, *History*, 2:350, 352–353 (25 February and 4 March 1669). A similar experiment was proposed by Papin, ibid., 4:524–525 (9 February 1687).

36. Ibid., 3:227 (8 July 1675).

37. S. J. Rigaud, *Correspondence of Scientific Men of the Seventeenth Century* (Oxford, 1841), II: 588 (24 August 1674); Wallis here refers to his *Mechanica, sive de Motu* (London, 1669–1671) as a possible first discussion of the projectile's trajectory as "compounded of one accelerated, the other retarded. . . . And practical cannoneers, I am told, find the random of a bullet very different from the parabola, which [Galileo's] hypothesis doth estab-

lish." Cf. Hall and Hall, *Correspondence of Henry Oldenburg*, XI: 108–109 (15 October 1674): "The line of projection in shootings, I have always suspected not to be a Parabola."

38. Birch, *History*, 3:398 (4 April 1678), 400 (18 April). All was not clear, however, for the parabola was to be vindicated again a few days later.

39. Ibid., 2:131–132 (4 December 1666); 344 (1 February 1669).

40. Ibid., 2:469 (9 February 1671); Sir William Petty, *The Discourse made before the Royal Society the 26th of November 1674 concerning the use of Duplicate Proportion in Sundry Important Particulars* (London, 1674), p. 1.

41. Birch, *History*, 3:115 (11 December 1673).

42. Ibid., p. 119 (18 December 1673), p. 136 (29 September 1674). For subsequent returns to the problem of organized experiment, see ibid., p. 309 (6 March 1676), and ibid., 4:6 (29 January 1680), 516 (5 January 1687). Hooke had been censured for failure to perform his duties properly as early as 14 November 1670 (ibid., 2:452). See also Michael Hunter, *Science and Society in Restoration England* (Cambridge: At the University Press, 1981), pp. 56–57.

43. Newton, *Correspondence* (Cambridge: At the University Press, 1975), 5:65–68 (21 September 1710), 70 (30 September 1710). J. Edleston, *Correspondence of Newton and Cotes* (London, 1850), p. 35.

44. Newton, *Correspondence*, 66:103–104 (24 March 1711). I have shortened the story: it was only as a result of Cotes's prodding that Newton finally explained (by the *vena contracta* effect) how the velocity of efflux calculated at $\sqrt{2gh}$—consistent with the height of the jet—was also consistent with a quantity of water flowing as though the velocity at the hole were $\sqrt{gh}$.

45. See A. Rupert Hall, "Newton and his Editors," *Notes and Records of the Royal Society* 29 (1974): 45; Whiteside, *Mathematical Papers of Isaac Newton* (Cambridge: At the University Press), 8:48–58.

46. Because in shooting over sloping ground the range is proportional to $\sin(2a-b)$, where a is the angle of elevation (above the horizontal) and b the angle of slope. Halley's papers are in *Phil. Trans.* XVI, no. 179 (January–February 1686): 3–21, and XIX, no. 216 (March–May 1695): 68–72.

47. See Hall, *Ballistics*, p. 94.

48. Charles Hutton, George Shaw and Richard Pearson, *The Philosophical Transactions*, abridged (London, 1809), 3:265.

49. E. F. MacPike, *Correspondence and Papers of Edmond Halley* (London: Taylor & Francis, 1932), p. 168.

50. Ibid.

51. H. Guerlac, "Vauban: The Impact of Science on War" in his *Essays and Papers in the History of Modern Science* (Baltimore and London: Johns Hopkins University Press, 1977), pp. 413–439.

52. Robert K. Merton, *Science, Technology and Society* (note 8), p. 204 (table). Morris Berman, *Social Change and Scientific Organization* (Ithaca: Cornell University Press, 1978), p. *xx*. I counted some fifty significant allusions to gunnery and allied matters in rereading Birch's *History*. The four volumes contain 2090 pages; reckoning five subject-allusions per page (surely an underestimation), we may discover altogether 10,000 subject

references, consistently counting each repetition of the same issue at subsequent meetings as a separate subject (thus one topic occurring ten times rates the same as ten topics occurring once). Merton's assessment produces a total of about 5,500 "problems" discussed at Royal Society meetings, presumably counting the total of repetitive allusions to the same "problem" as one: this would indicate a total number of all problems, ignoring repetitiveness, in excess of 10,000. According to my count, then, gunnery topics could not have been more than one percent, at the extreme, of all topics mentioned at meetings. But I am not convinced that the selection and classification of topics alluded to at meetings of the Royal Society can be made rigorous enough to provide more than a very rough guide to the significance of particular interests.

VI

NAUTICAL ASTRONOMY AND THE PROBLEM OF LONGITUDE

David W. Waters

Enabling a navigator to fix his position at sea scientifically in terms of latitude and longitude proved to be one of the greatest intellectual and technological challenges to Western man. The problem first came to the fore in the middle of the fifteenth century as a consequence of Prince Henry "the Navigator" of Portugal sending pilots out into the Atlantic to seek for islands and for a seaway to Guinea in West Africa.

The technique of scientifically measuring position at sea in a north-south direction in terms of latitude north or south of the equator was developed by nautical astronomy in the last two decades of the fifteenth century;[1] but measuring (as distinct from estimating) position at sea in an east-west direction in scientific terms of longitude east or west of a prime meridian remained a problem. The problem was to take some three centuries of effort to solve and, in the process, to stimulate scientific observation, inquiry, and invention in numerous fields.

The theory that longitude could be measured by measuring the difference in time between different places was well known to astronomers. It was also well known that longitude could be measured by observing and recording the time at which an eclipse or an occultation (the passage of the moon across a star) was seen to occur at two different places. The difference in time gave the measure of how far east or west one observer was from the other. Since in the course of a 24-hour day the sun appears to turn through 360°, conversion of the difference in time into degrees and minutes gave the longitude of one of the observers east or west of the meridian of the other. However, eclipses and occultations occur too infrequently to be of any use to a navigator. Some other means had to be found.[2] Meanwhile navigators continued to make the landfall after a long ocean passage by aiming for a point 200 to 300 miles east or west of the desired landfall and—when they reached its latitude, which they could measure to within $15' - 30'$ with a quadrant or sea astrolabe—by steering west (or east), "running down the latitude," until the landfall was sighted. As a landfall was chosen for its conspicuousness from seaward, the probability was that it would be sighted 15 to 30 miles distant (the latter distance being, on average, twice the visibility distance from a ship).

Thus, in 1577, Capt. Martin Frobisher, returning to England from his Northwest enterprise, described at length the use of the new nautical science in conjunction with traditional methods of pilotage; that is, latitude sailing by celestial observation when outside soundings, combined with lead and lookout when within soundings (the 100 fathoms line):

Having spend foure or five dayes in traverse of the seas with contrarye winde, making oure souther way good as neare as we could, to raise oure degrees to bring ourselves with the latitude of Sylley, we tooke the heighth the tenth of September, and founde ourselves in the latitude of (cypher) degrees and in ten minutes. . . . And upon Thursday the twelfth of September taking the height, we were on the latitude of (cypher) and a halfe, and reckoned ourselves not paste one hundred and fiftie leagues [450 miles] short of Sylley . . . then being in the height of Sylley . . . we kept our course east, to run in with the sleeve or channel so called, being our narrow seas, and reckoned as short of Sylley [one hundred and] twelve leagues . . . the fifteenth . . . we began to sounde with oure lead, and hadde grounde at sixty-one fathome depth . . . the seaventeenth . . . we . . . were shotte betweene Sylley and landesendes. . . .[3]

English navigators had learned their art with the aid of Portuguese navigators and from Spanish publications (themselves derived from Portuguese manuals) translated into English.[4] Although no printed fifteenth-century manual survives, the Portuguese had by the end of that century evolved a rule or table giving the distances to be steered on various courses to raise or to lay 1° of latitude and had prepared tables of the latitudes of important landfalls and places. By ca. 1500 they had added a scale of latitude to ocean charts so that by then all the basic elements for latitude sailing were in use.

In the course of the next two centuries many improvements were made in the art of navigation. Variation compasses and tables enabled the variation of the steering compass from true north to be measured—and allowed for—and charts to be constructed using true bearings. At the same time, the spiral nature of lines of direction (except those of the four cardinal points, N, S, E, and W) on a globe such as the earth was explained by the Portuguese cosmographer Pedro Nunes (1537), and from the mid 1590s, thanks to Edward Wright's tables of meridional parts, directional lines could be correctly represented on a chart. Altitude measuring instruments of greater accuracy, such as the cross-staff, and the backstaff invented by the English Capt. John Davis in the 1590s, supplanted the quadrant and sea astrolabe in the seventeenth century. The log-line and half minute glass, introduced by the English in the 1570s to measure instead of to estimate by eye alone the distance sailed, improved estimates of the speed of a vessel and so of the distances sailed. This was so particularly from the 1620s, when Edmund Gunter publicized the length of a degree, measured by triangulation by the Dutchman Snell in 1619. New mathematical tables, such as those of logarithms, and new calculating instruments, such as the sector and slide rule, made calculations easier and quicker from the early seventeenth century on.[5]

In the 1660s the length of a degree was measured even more accurately by the French astronomers Auzout and Picard. Nevertheless, in the last half of the eighteenth century, the navigator, regardless of how mathematically competent he might be—or how skilled an observer or how well equipped—was in practice as liable to err in his estimate of where he was in terms of longitude as were the Portuguese pioneers of the fifteenth century. In 1763 it could be stated on the highest authority—that of an astronomer experienced and skillful in the practice of navigation at sea—that:

Daily experience shews the wide uncertainty of a ship's place, as inferred from the common methods of keeping a reckoning, even in the hands of the ablest and most careful navigators. Five, ten or fifteen degrees [of longitude], are errors [of hundreds of miles] which no one can be sure he may not fall into in the course of long voyages.[6]

The navigator's only recourse was the same as it had been for the first ocean navigators, the Portuguese pilots of the fifteenth century: to "run down the latitude" from far to seaward, still timing with a sandglass the distance sailed, keeping a sharp lookout and sounding frequently long before the estimated time of approaching the continental shelf and the dreaded hazards of a shelving seabed (once soundings—bottom within the 100 fathoms [600 feet] line—were reached). This was still the best check on longitude next to sighting the intended landfall at the expected time.[7]

Why did it take so long to solve the problem of longitude at sea? It was not owing to lack of interest, or from a failure to realize the vital importance of finding a solution. It was the sheer practical—the theoretical, technological, and manipulative—difficulties that confronted successive generations of men in attempting to measure small intervals of time accurately—the intervals had to be small if they were to be useful for finding position at sea. The manipulative, manual problems of making precision measuring devices were difficult enough, but more fundamental were the theoretical and moral problems involved. This was because, for a long time, the key concepts were dangerous to the peace of mind and the physical well-being of many men. These men would be forced to think of the universe, of the earth and man's place in it, and of their creator differently from the way in which they had been brought up to believe. They would have to hold ideas radically different from, and have concepts of the world diametrically opposed to, those of all about them. Their thinking would have to be revolutionary.

The efficient cause of the arrival of this revolution in thought was the changing of the Christian calendar. The calendar had been erroneous; the feast days got out of step with the seasons. The church had always encouraged astronomy; indeed, until the last half of the fifteenth century, almost all astronomers were churchmen working either in a monastery or a university (here, however, some were Jews—as Abraham Zacuto of Spain, from whose astronomical tables of the 1470s the first Portuguese nautical almanacs were compiled). Successive generations of astronomers kept the Christian calendar, which had been inherited from the classical world, up to date. The calendar instituted by Julius Caesar depended on the accurate measurement of time, of the length of the so-called tropical year. This length was taken to be 365¼ days. Unfortunately, as this was slightly longer than the true solar year, an error of one day was accumulated every 120 years. By A.D. 1500 the error amounted to 10 days and was seriously affecting the date of Easter; so the reform of the calendar had become a matter of practical religious, as well as agricultural, importance, and it finally began to be attempted.

In 1543 a Polish canon of the Church of Rome, Nicholas Copernicus (who had been invited many years before by the Pope to assist in the reform of the calendar but had declined), published on his deathbed a small book. Its introduction, not written by him, described it as simplifying the method of calculating the calendar. The book was *De Revolutionibus Orbium Coelestium.*

It had been axiomatic since classical times that the heavenly bodies moved around the stationary, central earth, and because a circle was the aesthetically perfect shape, that they did so in circles. Copernicus claimed, with supporting arguments, that the cosmos could be represented better mathematically if the sun were considered stationary and central and the earth and planets were considered as revolving around the sun in the course of a year—with the earth, at the same time, rotating around its polar axis in a west-to-east direction. The enormity of this truly revolutionary concept of the universe—which had been considered and rejected by classical thinkers—was that it called in question the whole theological concept of man as the focus of God's attention. It relegated him to a minor position on the periphery of the world and so challenged the omniscience of the church and its divinely revealed wisdom—which the Church began to appreciate as Copernicanism spread.

Probably Copernicus was wise to delay publication of his book until the approach of death from natural causes. I shall put aside speculations about his motives, however, or about his thoughts on the ultimate intellectual and moral effects of his hypothesis. Suffice it to say that his hypothesis stimulated astronomical inquiry and measurement as well as religious controversy, for it raised, among many important questions, the fundamental ones of what moved the earth and planets and why they moved in apparently circular paths.

Through Galilean physics Copernicanism led to Newton's theory of universal gravitation and in 1666 to his testing of that theory by measuring the weight of the moon. This tower-ing intellectual feat was ultimately to make practicable, in conjunction with his method of fluxions (and the later differ-ential and integral calculus of Leibnitz for calculating the complex motions of the moon), one method of finding longi-tude at sea—the method based on lunar distances—though it was to take another hundred years for this to be accomplished.

The astronomer Johann Werner, in the 1514 edition of Ptolemy's maps (first published in the 1470s), first put forward in print the proposal that measurement of the angular dis-tance between the moon and the sun or a star could be used to find longitude.[8] This was an extension of the principle of observing an eclipse or the occultation of a star, when the angle of separation is zero. Werner proposed measuring the angle of separation between the moon and the sun or a carefully selected zodiacal star and comparing the local time at which that angle was measured by an observer at one place with the predicted time at which that same angle was to be seen at an observatory in a different place; then, the difference in time would give the longitude of the observer from the meridian passing through the observatory. Such observations could be made far more frequently than could observations that de-pended on eclipses or occultations. Clearly, this method as-sumed that the stars' positions are precisely known and cor-rectly catalogued, that they do not change perceptibly, and that the motions of the moon and sun have been analyzed accurately enough to enable them to be precisely predicted for months—indeed, for seamen making long voyages, years—ahead.

How precise is precisely? Consider the progress of the

moon. Against the background of the stars it circles the earth in the course of a month, moves through 360° in about 30 days, which is a movement from west to east of roughly 12° in a 24-hour day. In 2 hours it moves through 1° of arc, that is, 60' of arc in 120 minutes of time, or 1' of arc in every 2 minutes of time. One minute of arc is the limit of the resolving power of the unaided human eye; few men can observe as precisely; a limit of 3'−5' of arc is more usual.[9] The sun's, moon's, and stars' positions must therefore be predicted to within 1' of arc for observations of difference of longitude to be accurate to within 2 minutes of time. Time is kept by the sun. The sun (apparently) circles the earth through 360° from east to west in 24 hours, or through 30° in 2 hours. Compared with the moon's 1° west-to-east motion in 2 hours, this is 30 times as fast; in fact, in 2 minutes of time the sun moves through ½° of longitude, which, on the equator, is equivalent to 30 miles. We have thus quantified accuracy in relation to the navigator's needs and the astronomer's capacities, the precision on which they must base their predictions to enable seamen to use lunar distances: the astronomer must predict to within 1' of arc and the navigator observe to within 1' of arc. Then, if these errors do not cancel out, the navigator at sea should fix his longitude to within 1°—60 sea miles.

However, there was an alternative method of measuring longitude at sea, first put forward in 1530 by Gemma Frisius.[10] Weight-driven clocks were developed in Europe early in the fourteenth century. The first astronomer known to use a clock to time an observation was Bernard Walther, in 1484.[11] Gemma Frisius proposed that, spring-driven and therefore portable clocks having recently been developed, a clock set to the time at the place of departure should be carried during a journey. By this means, by comparing the local time with the clock's time, longitude could be determined from the difference between the times. As a voyage across the Atlantic could easily take 60 days, this meant that a sea clock would have to keep time to within 2 seconds a day. Unfortunately, clocks erred by 10 minutes a day; so a 300-fold increase in timekeeping accuracy was required, and it had to be maintained in all climates, from the poles to the equator, in calm and tempest, in a ship at sea.[12]

In short, neither method offered an easy solution, and as

the generations passed, finding longitude at sea came to be regarded as just as impracticable as squaring the circle, devising perpetual motion, or creating the philosopher's stone. Nevertheless, looking back down the corridors of time one perceives a positive change in the intellectual interests of learned men. As the seventeenth century approached, the natural philosopher replaced the cosmographer, and precision observation and measurement of phenomena became preoccupations, together with rational speculation based on experimentation and observation and the search for mathematical order.

The great breakthrough occurred in the year 1609, a date-peg for the start of the Scientific Revolution. In that year Galileo Galilei, a confirmed Copernican, first turned a telescope to the skies; the world has never been the same since, in fact or in imagination. He gave man the unprecedented ability to increase his powers of optical discrimination many fold. His inquisitive action enabled indistinct remote celestial objects, such as the moon and the planet Jupiter, to be brought close for detailed scrutiny.[13] Within a lifetime, minute angular differences between celestial bodies, and their motions in the skies, could be measured to within seconds of arc, thanks to Jean Picard aligning a telescope with a graticule to the sight bar of his quadrant in 1667, and to Adrian Auzout devising the attachment of an optical micrometer to the eyepiece.[14]

In January 1610 Galileo discovered the first four satellites of Jupiter with his telescope. He began recording the times of their frequent eclipses as they circled around the planet. Almost immediately he perceived that, as the ecclipses occur at the same absolute instant of time, this might offer a means of measuring longitude at sea. His protracted attempts to make this method practicable failed, but in 1668 Giovanni Domenico Cassini published an almanac giving the times of the eclipses of the satellites, and thenceforward, though of no practical use at sea because of the difficulties of observation, astronomers had a relatively frequent method of determining longitude on land to within one to three miles.[15] At last, the coastline of a country could be laid down with scientific accuracy on a chart, as Louis XIV discovered to his annoyance. When the results of the 1681 survey of France authorized by him were published, Ushant and the Brittany peninsula were moved back eastward on the chart, whose prime meridian passed through Paris, over 1½° of longitude.[16]

The year 1609 is also remarkable as the starting point of the Scientific Revolution because it was the year in which Johann Kepler published his first two laws of planetary motion derived from Tycho Brahe's obsessively accurate observations of, in particular, the planet Mars. Kepler noticed that, when collated, Tycho's observations of Mars differed by 8' of arc from the predictions in the tables. As Tycho's errors were no greater than 1' of arc, Kepler concluded that the orbit could not be circular and arrived finally at his two epoch-making laws: "the orbit of a planet is an ellipse with the sun situated at a focus," and "the radius vector, joining the sun to a planet, sweeps out equal areas in equal times." His third law followed in 1619: "the squares of the periodic times of any two planets are proportional to the cubes of the major axes of their orbits." He had found for the first time the paths followed by the planets, their varying speeds of travel, and their respective distances from the sun. He had thus established the plan of the solar system based on measurement and geometry rather than on traditional aesthetics and empiricism. His *Rudolphine Tables* of 1627 predicted phenomena upon inviolable natural laws and rigorous geometry and included Brahe's unprecedentedly accurate positions of the so-called fixed stars.

But why do the planets revolve around the sun, the moon around the earth? Why do they move in ellipses, not circles? Why do they not shoot off in straight lines? What mysterious force impels them? In 1666 Isaac Newton "began to think of gravity extending to the moon" and by a simple calculation quickly satisfied himself that the centripetal action of the moon's weight counterbalanced the planet's centrifugal tendency, thus conforming to the laws of the fall of bodies and Galileo's mechanics as governed by the principle of inertia.[17] It was weight which held the moon circling the earth.

Newton's full theory of universal gravitation was not worked out and published until 1687; and it gained acceptance, particularly on the Continent, where Cartesianism prevailed, only gradually.[18] Nevertheless, gravity was the key to predicting with increasing refinement the motions of the

moon. The increasingly accurate optical angle-measuring instruments and timekeepers revealed these motions to be ever
more complex because of gravity interacting upon the earth,
moon, sun, and planets. When methods of calculation were
developed commensurate with the complexities of the factors
involved, the first lunar tables suitable for fixing a ship's position at sea to within 1° of longitude became practicable. They
were compiled by Tobias Mayer in 1753.[19]

Gravity was involved in the method of determining longitude by using mechanical clocks, as well as in the lunar distance
method. Once again we go back to Galileo, this time to his
discovery of the isochronism of the pendulum, to whose value
as a controller of clocks he drew attention in 1637.[20] A clock
measures, in effect, the rotation of the earth, which it represents with an escape wheel whose rotation it records in hours
and minutes. In the seventeenth century the wheel was driven
by a weight or spring and allowed to escape at regular intervals—of, say, one second—by an escape mechanism. The
mechanism was made up of a verge or shaft with two pallets
that alternately locked and released the escape wheel so that it
rotated in steps of 1 second. The movements of the verge were
controlled by a foliot—a horizontal beam with weights on
either end—which swung to and fro under the impulses given
to it by the pallets. It had no natural period of oscillation,
however, and so was erratic. A pendulum, because it is controlled by gravity, has a natural regular motion—it is isochronous. It oscillates in equal spaces of time almost irrespective of the arc of swing or the weight of the bob. Christiaan
Huygens, it is generally conceded, invented the commercially
successful pendulum clock in 1657 and, of equal importance,
the spiral balance spring, which is also isochronous, in 1675.
This made accurate portable clocks and watches practicable.
Tested at sea in the 1660s and 1670s, however, Huygens's sea
clocks proved useless owing to "the motion of a ship, the
variation of Heat and Cold, Wet and Dry and the Difference
of gravity in different Latitudes," as Sir Isaac Newton was to
explain to the British Parliament that in 1714 was considering
the problem of longitude finding at sea.[21]

Just as Louis XIV had founded an observatory at Paris in
1667 to establish accurate maps of his domains, so in 1675

Charles II, hearing that the recorded positions of the fixed stars—and the tables predicting the motions of the sun, moon, and planets—were so erroneous as to be useless to seamen for determining longitude, declared "he must have them anew observed, examined and corrected, for the use of his seamen."[22] He thereupon appointed John Flamsteed his "astronomical observator" and built an observatory for him designed by Sir Christopher Wren, himself an accomplished astronomer in the royal park at Greenwich, "in order to the finding out of the longitude of places for perfecting navigation and astronomy."[23]

It had been Edward Wright, the mathematician-navigator, who in his *Certaine Errors in Navigation Detected and Corrected* of 1599 had declared that the solution of the longitude problem was beyond the ability of any one man and that it would be solved only when governments and princes encouraged individuals and provided ways and means. In France it was Louis XIV's minister Colbert who, seeing clearly that the basis of commerce was accurate land maps and sea charts and that the basis of these was astronomy, had created the Académie royale des sciences in 1666 and l'Observatoire de Paris in 1667, where many of the most brilliant scientists on the Continent were equipped and liberally financed to pursue science for the benefit of the state and of mankind in general.[24] From hence came the cartographical achievements of Auzout and Picard, which were based on their scientific measurement of the length of a degree on the meridian of Paris and used the method of triangulation devised by Gemma Frisius in the 1530s—and the telescopic sights and micrometers they themselves developed from the Dutch optical instrument makers and Galileo's telescopic inventions of 1609. And from hence also came Cassini's improved tables of the eclipses of Jupiter's satellites; the annual publication of *Connaissance des temps* from 1695; Huygens's improved timekeepers of the 1660s and 1670s; the measurement of gravity and the earth's shape; and the discovery by Roemer (a Dane) of the finite speed of light (1672–1676), to cite only the most notable men and achievements of l'Académie.[25]

In England Charles II's royal charter of 1662 establishing the Royal Society gave scientific inquiry a less formal structure but a widespread and penetrating influence—through (from 1665) the *Philosophical Transactions*; the inventive genius of the society's curator of experiments, Robert Hooke; and the diplomacy of one of its first secretaries, Henry Oldenburg. It was because of the influence of some of its Fellows that Charles II created the Royal Observatory at Greenwich and because Charles "best understood the business of the sea of any prince the world ever had" that successive Astronomers Royal were paid to devote their intelligence and energy undeviatingly to the problem of determining longitude at sea.[26] Until the middle of the nineteenth century, in fact, this was virtually their sole function and concern. This was a cardinal reason why it was in Britain that the two practicable methods were ultimately developed. In France, where they might have been developed earlier, the necessary continuity and singleness of purpose were lacking. Moreover, from 1714 the English Parliament offered an award of an extraordinary amount, £20,000—equivalent to more than $400,000 today—to anyone who devised a method of general utility in determining longitude at sea to within 30 miles, with lesser sums offered for greater distances. The Longitude Act, which established the prize, also set up a Board of Longitude consisting of men competent to assess solutions submitted, make grants for research, and recommend awards.[27] Thus, in Britain the problem was given not only continuity of research through the Astronomer Royal's charter but also great incentives by the Longitude Act.

In 1714, when the Longitude Act was passed, there was no observing instrument of suitable accuracy available for use at sea. In 1731 John Hadley, a vice-president of the Royal Society, who had been much interested in making reflecting telescopes (inspired by Isaac Newton's invention in 1670 of one that, because of its short focal length, was thought might be astronomically useful at sea), described a reflecting quadrant. This worked on the principle of double reflection, whereby the doubly reflected image of a heavenly body could be seen in the eyepiece of the viewing telescope simultaneously with the horizon. It was also capable of adjustment so that instrumental errors would be corrected.[28] Tested at sea in

1732, it enabled observations to be made to within 1′ of arc irrespective of the motion of the ship.[29]

This put the problem of longitude firmly back into the hands of the atronomers and horologists, for, with this instrument, the navigator could at last find his ship's local time at sea by an accurate observation of the altitude of the sun or a star and could measure a lunar distance with sufficient accuracy. What he needed now were lunar tables as accurate or a marine timekeeper of precision. However, before Hadley's accurate quadrant could be used to measure lunar distances, three corrections had to be made to the observations of the sun, moon, or star. First, there had to be a correction for the height of the eye above sea level, known as dip. For this, accurate tables existed based on geometry and the accurate measurements of the size of the earth made by the astronomers of l'Académie royale. Second, there had to be a correction for refraction, the bending effect of the atmosphere upon rays of light which increases the apparent altitude of heavenly bodies, being greatest near the horizon and zero at the zenith: thus, at 10° above the horizon the increase is 5′ of arc and at 45° it is 1′ of arc. By 1720 James Bradley, later to become the third Astronomer Royal, had produced—thanks to the recent invention of the barometer and the thermometer, with which atmospheric pressure and temperature could be measured—a table that was more than adequate for such corrections. The third correction that had to be made was for parallax. The moon's apparent position among the stars changes as a result of the observer being on the surface of the earth and not at its center. The resulting parallax can be nearly 1° of arc, depending on whether the moon is seen high or low in the sky. It was therefore necessary to know the ratio of the distance of the moon to the radius of the earth to within 1 part in 60, if the angular distance of the moon from a star was to be measured to within 1′ of arc—and this was known.

What of the moon and its predicted motions? John Flamsteed had pointed out in 1675 that the current tables, all derived from Kepler's *Rudolphine Tables*, had errors of up to 10′ of arc. Newton's *Mathematical Principles of Natural Philosophy*, to translate the Latin title of his *Principia* of 1687, provided a sound basis for understanding the moon's irregularities. He showed that the motion of the moon around the earth could be explained by gravitational attraction between two bodies and that the major irregularities of this motion were due to the gravitational attraction of the sun. He showed also that, if the sun did not affect it, the moon would move around the earth in an ellipse, following Kepler's laws of one body moving around another. But the sun does affect the moon, and he showed that the major effect is to make the moon's axes of orbit rotate slowly every 8.85 years and, at the same time, to make the positions in the skies where this orbit intersects the plane of the earth's orbit move slowly through an 18.6-year cycle. These effects also accounted for three naked-eye effects: (1) variation, the wobble due to the sun and earth pulling together and then in opposition during a month, causing a 39' variation in the moon's position in the sky; (2) eviction, the sun's gravitational effect upon the moon's elliptical orbit in a month, amounting to up to 76' of arc displacement in a month; (3) annual inequality, discovered by Kepler, and now explained as caused by the sun's pull on the moon varying because of the ellipticity of the earth's orbit, which displaces the moon by about 11' of arc. Besides these major effects, there were others caused by the earth's shape not being spherical (due to its axial rotation it bulges at the equator, as a consequence of which gravity changes with latitude) and by attractive effects of the other planets on the moon.

Newton had predicted the earth's precession, the slow circular motion of its inclined axes completed in a cycle of about 26,000 years, and that it would not be perfectly circular because the axes of the moon's orbit rotate about the earth's once every 18.6 years. This effect, known as nutation, was discovered and explained by Bradley in 1747 as a result of observations made by him with a zenith sector of unprecedented accuracy made by George Graham, F.R.S. Bradley found that superimposed on the circular motion of the earth's axis were small waves of about 9' of arc with a period of 18.6 years; so Newton was vindicated and corrections for this could now be soundly based.[30] The theory of gravitation began to be indisputable.

What about the positions of the so-called fixed stars?

When Flamsteed became Astronomer Royal, Brahe's catalogue of the stars was still the standard; so it was the situation "that we were so far from having the places of the fixed stars true, that the Tychonic catalogues often erred ten minutes or more; that they were uncertain to three or four minutes, by reason that Tycho assumed a faulty obliquity of the ecliptic and had employed only plain sights in his observations," to use Flamsteed's own words.[31] As a result of the use of telescopes with micrometer eyepieces for measuring the positions of the stars, and of the pendulum clock for timing their transits, Flamsteed was able to observe their positions to within 10″ of arc, a 600-fold increase in accuracy; but his great catalogue, *Historia Coelestis Britannicae*, was only published in 1725, five years after his death.

In Paris in 1672 Cassini had organized a measurement of the planet Mars, from which, by Kepler's laws, the size of the earth's orbit was worked out and a reasonable estimate made of the earth's speed in its orbit. Three years later at Paris, Roemer, by observing and timing the eclipses of Jupiter's satellites, discovered and measured the velocity of light.[32] In 1727, thanks to Roemer's discovery, Bradley discovered the aberration of light—the shift in a star's position of up to 20″ of arc in a year—and was able to explain the shift as caused by light traveling at 300,000 kilometers per second and the earth moving around its orbit at a speed of 30 kilometers per second, so that the apparent direction of a star varies during the year from zero up to 20″ of arc. As a result of these discoveries, by 1753 many bright star positions were known to within 5″ of arc.

At this point Tobias Mayer, a German astronomer who never saw the sea, enters the story of longitude at sea. He was dedicated to determining accurately the latitude and longitude of places in order to improve cartography. The problem of longitude particularly concerned him. After having read the Swiss mathematician Leonard Euler's paper on analysis and mechanics in 1751, he was able to investigate the moon's motion more rigorously and to calculate a large number of new inequalities in the moon's motion due to the spheroidal shape—the equatorial bulge—of the earth.[33] He was able to confirm the opinion of the second Astronomer Royal, Edmond Halley, that there was a secular acceleration of the moon's mean motion owing to the influence of the earth's spheroidal shape, and to include these various improvements in his "New tables of the motions of the sun and moon" published in Göttingen in 1753. In the next few months he compared his predictions with reliable measurements of the moon's position made by James Bradley, now the third Astronomer Royal, during the years 1743–1745 with a mural quadrant by George Graham, F.R.S., which agreed to within 1' of arc. He was then able both to improve further the accuracy of his lunar tables and to establish that errors in observed star positions were a major reason for the discrepancies of 15' in lunar predictions being found by astronomers such as Euler. He was also able to discover errors in contemporary star catalogues which, if ignored, would have produced apparent errors of several minutes of arc in the moon's predicted positions.

At last, on the prompting of Euler, in January 1755 Mayer submitted to Admiral Lord Anson, president of the Board of Longitude, his proposals and tables for finding longitude at sea by the new method devised by him which required only a single observation of the moon and a zodiacal star. They were reported on favorably by Bradley in 1756 and 1760, after they had been tested out successfully at sea by Capt. John Campbell of the Royal Navy. After discarding Mayer's reflecting circle as unhandy, he, with the instrument maker John Bird, devised the sextant for taking the lunar distances more accurately and more often than was possible with a Hadley's quadrant. The board deferred judgment, however, because a competitor was in the field.[34]

Meanwhile, in 1761 a young astronomer, Nevil Maskelyne, was sent to St. Helena by the Royal Society to observe the transit of the planet Venus across the face of the sun; this had been predicted by Halley many years before and proposed by him as an improved method of measuring the size of the solar system. Maskelyne took with him Mayer's manuscript tables of 1754, a Hadley's reflecting quadrant of 20-inch radius made by John Bird, and a pocket watch. Thus equipped, he took lunar distances at sea, by which after some four hours of calculation he was able to determine longitude to within about one degree and often within half a degree. On his return he prepared *The British Mariner's Guide*, wherein he presented directions and tables (corrected for "the meridian of the Royal Observatory at Greenwich") derived from Mayer's printed tables of 1754. It was published in 1763 and used (by the purser, a keen amateur astronomer) during Capt. Wallis's voyage around the world (1766–1768) to fix the position of his ship, H.M.S. *Dolphin*, and the newly discovered island of Tahiti in the South Pacific.[35]

Before he died in February 1762 Mayer knew his proposed method worked at sea, for a Dane, Carsten Niebuhr, whom he had trained to the purpose in Göttingen, sent him the results of lunar distances taken by him at sea. This persuaded the dying Mayer to get his wife to send to the Board of Longitude his latest reissued tables. This his widow did in 1763. Two years later Parliament awarded her £3,000 and awarded to Euler, for the basic equations used to calculate Mayer's tables, an unsolicited £300.[36] Longitude finding to within 1° was at last practicable at sea—but only for a navigator who was an exceptionally able mathematician.

The competitor for the longitude prize in the 1760s whose solution deferred the Board of Longitude's decision to recommend an award to Mayer, was a Yorkshireman, John Harrison. He had started life as a carpenter and become an extraordinarily accomplished clock- and watchmaker. As more is learned of him and his handiworks, he is more clearly seen to have been a genius. He was obsessed, and rightly, with the two chief causes of irregularity in a mechanical timekeeper. These are, changes of temperature—which cause a pendulum to lengthen or shorten and (in a watch) the balance to change its size and the balance-spring its elasticity—and excessive friction, arising from inefficiency in the escapement, gearing, and pivots. With his brother James he devised a bi-metallic grid-iron pendulum that compensated for changes of temperature, and an almost frictionless escapement, bearings, and pivots. Then, in 1730, determined now to win the longitude prize, he went to London with plans for a sea clock of revolutionary design incorporating these improvements.[37] Encouraged by Halley—the Astronomer Royal—and George Graham, the leading instrument- and clockmaker in London, to whom he showed his plans, he returned home to make the longitude machine, or sea clock. It was completed and tested at sea in 1735 as "Public Encouragement, In order to a thorough Tryal and Improvement of the severall Contrivances, for preventing those Irregularityes in time, that naturally arise from the different degrees of Hot and Cold, a moist and drye Temperature of the Air, and the Various Agitations of the ship," as five eminent Fellows of the Royal Society certified in 1735.[38] H.1 (as it is known) performed sufficiently satisfactorily for Harrison to pursue his work with the encouragement and financial support of the Board of Longitude.

He then made three more marine timekeepers, each of novel design, of which the last, H.4 (radically different in appearance for it was a watch), was to be tested twice at sea. H.2 was completed in 1739, H.3 in 1757, and H.4 in 1759.[39] Since the mid-1750s, therefore, Harrison's work had been almost neck and neck with Mayer's lunar tables in competing for the longitude prize; hence the Board of Longitude's circumspection in awarding any prize. In 1761, H.4 was tested on a voyage to Jamaica which lasted 81 days and, allowing for the timekeeper's rate, resulted in a cumulative error of 5.1 seconds of time, or, in the latitude of Jamaica, of just over one mile in longitude. It was incredible. The longitude of Kingston, Jamaica, had been found some years previously by an eclipse of the moon and a transit of Mercury, so it was well established. Nevertheless, it was decided that H.4 had to be tested again. This time the test was a voyage to Barbados and it took place in 1764. The longitude of the island was established by Nevil Maskelyne and an assistant by observations of eclipses of the first satellite of Jupiter, and it was Maskelyne who checked the timekeeping of H.4 on its arrival by equal altitude observations. It was perhaps understandable that John Harrison and his son William (who actually accompanied H.4 on the voyage) should regard Maskelyne with suspicion. *The British Mariner's Guide* had been published only the year before, and on his voyage out (as on his return) Maskelyne was directed by the Board of Longitude to "make observations of the Moon's motions, and to try the accuracy of the said (Mayer's) tables," which he did with great success, a fact he not unnaturally boasted abroad in Barbados.[40]

By Harrison's reckoning, over the round-trip voyage of 156 days H.4 lost only 54 seconds, or a mere ⅓ second a day. The official result was an average error of only 39.2 seconds of time, equal in the latitude of Barbados (16°N) to 9.8 miles and thus well within the 30 miles qualifying for the maximum prize. Of this Harrison now received half, £10,000. Ultimately, after appealing to George III, another scientifically minded king, Harrison was awarded a further £8,750, he being then of sound mind but 80 years old. He died three years later.[41] By then Capt. James Cook had returned from his second Pacific voyage (1772–1775) and had reported enthusiastically on the reliability and accuracy in Antarctic cold and tropic heat, in calm and gale, of K.1, Larcum Kendall's copy of H.4.[42] The longitude machine, Harrison's dream of the sea clock for longitude, had become reality. Henceforth, longitude at sea could be determined to within a mile or so using chronometers.[43]

Meanwhile, until commercial production of chronometers provided a sufficiency (Harrison made five in his lifetime), lunars were the chief means of general utility—particularly as Maskelyne, when he became astronomer royal in 1765, immediately undertook the official annual publication of the *Nautical Almanac*. The first, published for the year 1767, appeared at the end of 1766 together with *Tables Requisite* for taking lunar distances. It was authorized by the commissioners of longitude as "a Work which must contribute to the Improvement of Astronomy, Geography and Navigation." The tables were simplified for the mariner's use "so that finding the Longitude by the Help of the Ephemeris is now in a Manner reduced to the Computation of Time, an Operation equal to that of an Azimuth [for determining compass variation, with which mariners were familiar] and the Correction of the Distance on account of Refraction and Parallax, which is also rendered very easy." This is Maskelyne speaking in his preface; he rightly and generously adds, "All the Calculations of the Ephemeris relating to the Sun and Moon were made from Mr Mayer's last manuscript Tables, received by the Board of Longitude after his Decease. . . . The Calculations of the Planets were made from Dr Halley's Tables."[44]

So the long struggle was over at last. From the preoccupations of longitude at sea and of nautical astronomy, astronomers were now about to direct their thoughts and their telescopes to the structure of the heavenly bodies. Physical astronomy may be said to have been born with the labor pains of positional, particularly nautical, astronomy—it having been begotten by the problem of "guiding a ship engulfed, where only water and heaven may be seen."[45]

Jesse Ramsden's invention of the scale dividing machine in the 1770s made generally available to seamen small, handy sextants and quadrants of unprecedented standards of accuracy, for astronomical observations for longitude.[46] The commercial production of marine chronometers from the 1780s, notably by John Arnold and Thomas Earnshaw, made Gemma Frisius's 1530 proposal to find longitude with a clock one of general and increasing utility at sea. Europe had already passed from the Scientific Revolution into the Industrial Revolution, which had been, and would continue to be, pioneered worldwide by virtue of the ability of seamen to find "the so much desired longitude of places" at sea. The exploration—and exploitation—so long deferred, of one third of the surface of the world, the Pacific Ocean, could proceed at last: "California, here I come."

As a memorial to the scientists, the instrument makers, and the seamen whose endeavours over the centuries made this possible, the old Royal Observatory, where so much was done to this end, stands proudly and yet benignly on the crest of its hill overlooking the River Thames.[47] It stands as a visible symbol of a triumph of determination and a victory for freedom of thought, and as a monument to those men "who go down to the sea in ships and occupy their business in great waters," who by their knowledge of the stars' places and the motions of the sun, moon, and planets come at last in safety unto a haven where "Port after stormie seas doth greatly please."[48]

NOTES

1. The history of this achievement has been studied in depth by Portuguese scholars, in particular over the last century. During the last seventy years many invaluable facsimiles of early sixteenth-century navigational

works have been published in whole or in part. Notable are J. Bensaúde, *L'Astronomie nautique au Portugal à l'époque des grandes découvertes* (Berne: M. Dreschel, 1912), and his seven facsimile volumes, detailed in D. W. Waters, *Science and the Techniques of Navigation in the Renaissance*, National Maritime Museum Monograph no. 19 (London, 1976; 2d ed., 1980), p. 9 n. 6; L. Pereira da Silva, *Obras completas*, 3 vols. (Lisboa, 1943–1946); L. M. de Albuquerque, *Curso de história da naútica* (Coimbra: Livraria Almedina, 1972); J. I. de Brito Rebello, ed., *Livrò de Marinharia* (Lisboa, 1913); A. Cortesão, *History of Portuguese Cartography*, 2 vols. (Coimbra: Junta de Investigações do Ultramar, 1969–1971); A. Cortesão and A. Teixeira da Mota, *Portugaliae Monumenta Cartographica*, 6 vols. (Coimbra, 1960–1962); R. H. Major, *The Life of Prince Henry of Portugal Surnamed the Navigator* (London, 1868); E. G. R. Taylor, *The Haven-Finding Art* (London: Hollis and Carter, 1968) which does not cite sources; and "The Navigating Manual of Columbus," *Jour. [Roy.] Institute of Navigation* (London), vol. 5, no. 1 (January 1952), which is a key synthesis of fifteenth-century nautical science. C. H. Haring, *Trade and Navigation between Spain and the Indies in the Time of the Hapsburgs* (Cambridge, Mass.: Harvard University Press 1918; reprinted, Peter Smith, Gloucester, Mass.: 1966), gives a masterly account of navigational developments in Spain in the sixteenth century. Waters, *Science and the Technique of Navigation*, details many other titles and their particular relevance; for early Portuguese sailing directions see D. W. Waters, *The Rutters of the Sea* (New Haven, Conn.: Yale University Press, 1968).

2. For a full study of the problem see D. Howse, *Greenwich Time and the Discovery of the Longitude* (Oxford: Oxford University Press, 1980). It is well documented.

3. D. W. Waters, *Science and the Techniques of Navigation in the Renaissance*, National Maritime Museum Monograph no. 19 (London, 1980), p. 29, where a more extensive account is given. This gives many Renaissance sources. For Frobisher's voyages see D. W. Waters, n. 5 below, p. 145 f.

4. Waters, *Science and the Techniques of Navigation*, esp. pp. 7–19.

5. D. W. Waters, *The Art of Navigation in England in Elizabethan and Early Stuart Times* (London: Hollis and Carter, 1958; 2d ed., Greenwich: National Maritime Museum, 1978), passim, where these are dealt with exhaustively and with a comprehensive bibliography.

6. N. Maskelyne, *The British Mariner's Guide* (London, 1763), preface, p. i.

7. Thus, in 1768, when sending Lt. (as he then was) James Cook out to Tahiti in 1768 to observe the transit of Venus, the lords of the admiralty instructed him "to put to sea and proceed round Cape Horn to Port Royal Harbour in King Georges Island [Tahiti] situated in 17 degrees and 30 minutes south latitude and 150 degrees of longitude west of the meridian of the Royal Observatory at Greenwich . . . taking care however to fall into the Parallel of King Georges Island at least 120 leagues [360 miles] to the east of it." J. C. Beaglehole, ed., *The Journals of Captain James Cook on his Voyages of Discovery*, 4 vols. and a portfolio, 1, *The Voyage of the Endeavour* (Cambridge: At the University Press, 1968): cclxxix–cclxxxi. This is probably the first sailing instruction to give the longitude as well as the latitude of a captain's

destination. It had been established at the time of the discovery of Tahiti by Capt. Wallis in 1767. He had an amateur astronomer on board who had volunteered for the voyage and determined the longitude of Tahiti (and elsewhere) astronomically. Despite Cook having an experienced astronomer on board who could find longitude at sea by lunar distance (as, indeed, Cook could and did), he was ordered to run down the latititude to make Tahiti.

8. J. Werner, *In hoc opere haec cotintentur Nova translatio primi libri geographicae Cl'Ptolomei* . . . , (Nuremberg, 1514), in his commentary on Chap IV.

9. See, C. Singer et al., *A History of Technology*, Vol. III (Oxford: Clarendon Press, 1957); D. J. Price, "Precision Instruments: To 1500," p. 583, Table I, "Orders of Accuracy in Astronomical Observation."

10. G. Frisius, *Gemma Phrysius de Principiis Astronomiae & Cosmographiae . . . vsv globi et eodem editi* (Antwerp, 1530), sigs. D_2 ^{V}–D_3 ^{r}. In 1553 he added a paragraph that specifically recommended the chronometer method for "sea-journeys." The passages are in translation in Howse, *Greenwich Time*. The first description of the method in English was given by W. Cuningham, in *The Cosmographical Glasse* (London: J. Daye, 1559), fol. 110, reproduced in Howse, *Greenwich Time*, p. 11.

11. D. de B. Beaver, "Bernard Walther: Innovator in Astronomical Observation," *Journal for the History of Astronomy* 1, pt. 1 (Feb. 1970): 40–41, quoting (in translation) from J. Schöner, ed., *Scripta clarissimi mathematici* (Norimbergae, 1544), fol. 50 R-v.

12. See F. A. B. Ward, *Time Measurement: Historical Review* (London: Science Museum, 1970), p. 8, fig. 2.

13. His *Siderius nuncius* was published in Venice in March 1610; it announced the discovery of the mountains of the moon, innumerable stars, and the (first four) satellites of Jupiter. It comprised a mere twenty-four leaves in octavo.

14. R. Taton, "Les origines et les débuts de l'observatoire de Paris," text of a paper presented at the Tercentenary of the Royal Observatory, Greenwich, 14–18 July 1975, p. 6g: "S'ils n'inventèrent pas le micromètre à fil, et s'ils ne furent pas les premiers à songer à adapter des lunettes aux instruments de mesure angulaire, en revanche les astronomes parisiens, en particulier Picard, Auzout et peut-être Roberval, furent bien les premiers à en faire en usage courant et à les adapter aux principales mesures astronomiques et géodésiques.[15]" His footnote 15 reads: "Cf. R. McKeoln, in *Physis*, XIII, 1971, p. 225–288 et XIV, 1972, p. 221–232; J. W. Olmsted, in *Isis*. XL, 1949, p. 213–225." Taton's paper is printed in *Vistas in Astronomy* (Oxford: Pergamon Press, 1976), 20: 65–71.

15. D. Cassini, *Ephemerides Bononiensis mediceorvm sydervm ex hypothesbvs, et tabu is lo: Dominici Cassini* . . . Bologna: Bononiae, 1668).

16. The map was reproduced in *Rec. de l'Acad. des Sc.*, To. VII, Pl. VIII, 430, and is reproduced in Howse, *Greenwich Time* (n. 2 above), and in Lloyd A. Brown, *The Story of Maps* (New York: Bonanza Books, 1949), facing p. 246.

17. A. Koestler, "The Newtonian Synthesis," in *The Sleepwalkers* (Lon-

don: Penguin Books, 1975, first published 1959), pp. 504–513 discusses this lucidly. Galileo's fundamental contributions are discussed by C. Singer, *A Short History of Scientific Ideas to 1900* (Oxford: Clarendon Press, 1962, first published 1959), pp. 230–236, notably Galileo's ideas in his *Dialogo sopra due massimi sistemi del mondo* (1638), which is available in a translation by Stillman Drake (Galileo Galilei, *Dialogue Concerning the Two Chief World Systems, Ptolemaic and Copernican* [Berkeley and Los Angeles: University of California Press, 1953]).

18. I. Newton, *Philosophiae naturalis principia mathematica* (Londini, 1687).

19. T. Mayer, "Novae tabulae motuum solis et lunae," *Commentarii Societatis Regiae Scientiarum*, Gottingensis, ii (1753), pp. 383–430, and N. Maskelyne, ed., *Tabulae motuum solis et lunae novae et correctae: auctore Tobias Mayeri* (Londini, 1770). The importance of Mayer's tables is developed in E. G. Forbes, *The Birth of Navigational Science*, National Maritime Museum Monograph no. 10 (London, 1974). The advances in horological and instrumental accuracy are well discussed and documented in D. Howse, *Greenwich Observatory*, vol. 3, *Buildings and Instruments* (London: Taylor and Francis, 1975).

20. Galileo, in a letter dated 6 June 1637 to Lorenzo Real, admiral of the Dutch East India Company. See S. A. Bedini, "Galileo Galilei and Time Measurement: A Re-examination of Pertinent Documents," *Physis* 5, fasc. 2 (1963): 145–165, especially p. 150. See also *Le Opere di Galileo Galilei* (Florence: Edizione Nazionale, 1909), 19:654.

21. *House of Commons Journal* 17 (11 June 1714).

22. F. Bailey, *An account of the Revd John Flamsteed the first astronomer Royal compiled from his own manuscripts* . . . (London, 1835), p. 38. The original ms. from which Flamsteed's words were taken by Bailey is in the Royal Greenwich Observatory, Herstmonceux, Sussex. Flamsteed also wrote: "When Charles II, King of England, was informed of these facts, he said the work must be carried out in a royal fashion. He certainly did not want his ship-owners and sailors to be deprived of any help the heavens could supply, whereby navigation could be made safer." J. Flamsteed, *Historia Coelestis Britannica*, Vol. III (London, 1725), prolegomena, p. 102. See D. Howse, *Francis Place and the Early History of the Greenwich Observatory* (New York: Science History Publications, 1975), where all Place's engravings of the Royal Observatory made shortly after its completion in 1675 are reproduced in facsimile.

23. E. G. Forbes, *Greenwich Observatory*, vol. 1, *Its Origins and Early History* (London: Taylor and Francis, 1975), and H. D. Howse, *Greenwich Observatory*, vol. 3, *Buildings and Instruments* (London: Taylor and Francis, 1975), should be consulted for the history of the observatory. J. A. Bennett, "Christopher Wren: Astronomy, Architecture, and the Mathematical Sciences," *Journal of the History of Astronomy* 6, pt. 3 (October 1975): 149–184, is a brilliant, exhaustively documented study of the architect/astronomer.

24. R. Taton, "Les origines et les débuts de l'observatoire de Paris," *Vistas in Astronomy* 20 (1976): 65–71.

25. S. Grillot, "Le problème des longitudes sur terre," *Vistas in Astronomy* 20 (1976): 81–84; J. Levy, "La création de La Connaissance des Temps,"

Vistas in Astronomy 20 (1976): 75−77. "La mise en évidence par Römer à la fin de 1675 et au cours de 1676 de la vitesse finie de la lumière est l'un des plus brillants exemples des résultats obtenus en quelques années par l'équipe d'excellents astronomes réunis autour de l'Observatoire de Paris. Mais, en les amenants à consacrer une partie croissante de leur activité à des opérations techniques de leves topographiques et de nivellement, le pouvoir publique de tutelle allait freiner ce remarquable effort d'observation et de recherche qui servira, au moins en partie, d'exemple aux très importants travaux qui seront dès lors menés à l'Observatoire de Greenwich" (R. Taton, "Les origines de l'observatoire de Paris," pp. 65−71 [p. 70]).

26. S. Pepys, *Naval Minutes*, ed. J. R. Tanner (London: Navy Records Society, 1926), pp. 71−72.

27. The Longitude Bill was 12 Ann. Cap. 15. as approved on 8 July 1714 "for providing a public reward for such person or persons as shall discover the longitude at sea." Its origins are well described and documented by S. M. Farrell, "William Whiston: The Longitude Man," *Vistas in Astronomy* 20 (1976): pp. 131−134.

28. J. Hadley, "The Description of a New Instrument for taking Angles," *Phil. Trans.*, no. 420 (1731), pp. 147−157.

29. J. Hadley, "An Account of Observations made on board the Chatham Yacht . . . ," *Phil. Trans.*, no. 423 (1732), pp. 341−356.

30. See E. G. Forbes, *Greenwich Observatory*, vol. 1 (n. 23 above) and Howse, *Greenwich Observatory* (n. 23 above) 3: 60−64 and figs. 54−57, for Bradley's work and zenith sector; also S. P. Rigaud, *Miscellaneous Works and Correspondence of the Rev. James Bradley, DD FRS* (Oxford, 1832).

31. Bailey, *Account of the Revd John Flamsteed, p. 20*.

32. "Dès l'année 1667, les astronomes de l'Académie, Auzout, Picard, Huygens et Roberval - Cassini et Römer participerent plus tard de réflexion analogue-avaient élaboré de vastes programmes d'observations et de recherches. . . . Les principaux documents originaux à étudier sont malheureusment très lacunaires." R. Taton, "Les origines de l'Observatoire de Paris," in *Vistas in Astronomy*, 20: 69−70. He then summarizes these sources and their locations (when not published).

33. E. G. Forbes, *The Euler-Mayer Correspondence (1751−1755)* (London: Macmillan, 1971), Letter I, pp. 33−36.

34. This is discussed, with detailed references, in Forbes, *Birth of Navigational Science* (n. 19 above), especially on pp. 6−9.

35. A. Carrington, ed., *The Discovery of Tahiti: A Journal . . . by . . . George Robertson* (London: Hakluyt Society, 1948), pp. *xxxix*, 216−220. Robertson was the master (the navigating officer) of the ship and was not able to take lunars. These (and other precise astronomical observations) were taken by the purser, John Harrison (no relation to the chronometer maker of that name), who was a keen amateur astronomer and had volunteered for the voyage, under the command of Captain Wallis, who did not understand lunars either. The facts about John Harrison, purser in the *Dolphin*, are contained in a letter by Captain Wallis to the earl of Egmont, 19 May 1768 (Shelburne MSS., vol. 75, pp. 435−445, William L. Clements Library, Uni-

versity of Michigan). Nevil Maskelyne's report of his lunars taken en voyage to St. Helena in 1761 are in *Phil. Trans.*, LII, pt. 1 (1762): pp. 558–577.

36. The relevant documents are indexed under "Payment of Reward of Tobias Mayer to his widow," in vol. 1 of the Board of Longitude papers (R. G. O. MSS; PRO Ref. 529, pp. 145–155). They constitute the basis of the article by E. G. Forbes, "Tobias Mayer's Lunar Tables," *Annals of Science* XXII (1966): 105–116. The award was made under the terms of Act 5 George III, Cap. xx. Euler's award was confirmed by the Board of Longitude on 25 May 1765. E. G. Forbes, "The Foundation and Early Development of the Nautical Almanac," *Jour. [Roy.] Inst. of Navigation* 18, no. 4 (Oct. 1965): 393.

37. Not 1727 as stated in *Memoir of a Trait in the Character of George III* . . . , by Johan Horrins . . . , (London, 1835), p. 215. (Johan Horrins is an anagram of the name of John Harrison's grandson.) Not 1728 as stated in the *Annual Register* for 1777, p. 25 of "Characters." H. Quill, *John Harrison: The Man Who Found Longitude* (London, 1966), p. 33.

38. The certificate, in the Monson Papers, Lincolnshire Archives Office, is transcribed in H. Quill, *John Harrison*, p. 42. Colonel Quill's biography is the only full-length study of John Harrison's life and is exhaustively researched, scrupulously annotated, and provided with a detailed bibliography. It does not detail the mechanisms of Harrison's various timekeepers—land and sea; for those of his marine timekeepers see n. 39 below.

39. The mechanical details of his marine timekeepers are described as follows: H.1, in the *Horological Journal* for December 1923, June 1951, and June, 1952, and in R. T. Gould, *The Marine Chronometer*, (London, 1926; reprinted 1960), pp. 40–47; H.2., in the *Horological Journal*, April 1953, and Gould, *Marine Chronometer*, p. 47; H.3, the *Horological Journal*, March, April, May, June, and July 1932, April 1953, and Gould, *Marine Chronometer*, p. 48–49; H.4, the *Horological Journal*, April 1953, December 1955, and Gould, *Marine Chronometer*, pp. 49–65; H.5, the *Horological Journal*, December 1955, pp. 804–809, and Gould, *Marine Chronometer*, p. 65. H.1, H.2, H.3 and H.4 are all on display in the Navigation Room, National Maritime Museum, Greenwich, where they are kept running. H.5 is in the Clockmakers Company Museum, Guildhall, London. An excellent brief account of Harrison's marine timekeepers is R. T. Gould, *John Harrison and His Timekeepers*, first published by the National Maritime Museum in 1958 and frequently reprinted. It was originally published in *The Mariner's Mirror*, April 1935.

40. Board of Longitude minutes, 4 August 1763. Scholars interested in studying the Board of Longitude papers will find invaluable E. G. Forbes, "Index of the Board of Longitude Papers at the Royal Greenwich Observatory," part 1, *Journal for the History of Astronomy* 1, pt. 2 (Aug. 1970): 169–179 (vols. I-XV); part 2, ibid., 2, pt. 1 (February 1971): 58–70 (Vols. XVI-XXIX); part 3, ibid., 2, pt. 2 (June 1971): 133–145 (Vols. XXX-LIV). The papers have been microfilmed and a copy is held by the New South Wales government; another (a Xerox) is held in the National Maritime Museum, Greenwich; microfilm copies of all or any of the 54 volumes can be obtained

on application to the Public Record Office, Kew, London. Two supple-
mentary sources are held in the P.R.O., namely: "Compilations of minutes
and orders relating to Board of Longitude," from 1713−1775 (P.R.O. Ref:
ADM 7/684), and "Papers relating to Longitude 1763−1819" (P.R.O. Ref.:
ADM 49/65).

41. John Harrison died at his house in Red Lion Square, London, and
was buried in Hampstead Church on 2 April 1776 (but the register errone-
ously records him as "Thomas" Harrison).

42. Captain Cook made a number of specific entries in the log of H.M.S.
Resolution referring to the replica of H.4 made by Larcum Kendall by order
of the Board of Longitude and taken by Cook on his second voyage to the
Pacific to give the chronometer a thorough testing by circumnavigating the
Antarctic. For example, "Mr. Kendel's Watch has exceeded the expectations
of its most zealous advocate and . . . has been our faithful guide through all
the vicissitudes of climates," recorded Cook on reaching the Cape of Good
Hope on his return voyage in March 1775. J. C. Beaglehole, ed., *The Journals
of Captain James Cook* (Cambridge, 1965), II: 660. Published for the Hakluyt
Society, London, Beaglehole's three volumes and portfolio of Cook's Pacific
journals (1769−81) and his life of Cook are fundamental and magisterial.
H. D. Howse and B. Hutchinson, *The Clocks and Watches of Captain James
Cook, 1769−1969* (reprinted from the four quarterly issues of *Antiquarian
Horology*, National Maritime Museum, Greenwich, 1969), should be consulted over the testing of the marine timekeeper in "remote parts" between
13 July 1772 and 29 July 1775, when Cook's *Resolution* departed from
Plymouth Sound and returned to Spithead, having traversed equatorial and
polar regions in the course of these three years. This study also discusses and
illustrates Captain Cook's pendulum clocks, the saga of the Shelton clocks,
his minor clocks and watches, and the Kendall watches and Arnold chronometers tested on his second and third (and fatal) voyage, when he explored the North Pacific and Arctic Ocean through the Bering Strait.

43. D. W. Waters, "Navigational Problems in Captain Cook's Day," in
Exploration in Alaska, Captain Cook Commemorative Lectures (Anchorage,
Alaska: Cook Inlet Historical Society, 1980), pp. 40−56, describes the
practice of navigation in Captain Cook's time (with sources) and discusses
the problem of the use of lunars and chronometers in the next decades. G. V.
Nockolds, "Early Timekeepers at Sea," *Antiquarian Horology*, Sept. and Dec.
1963, W. E. May, "How the Chronometer Went to Sea," *Antiquarian Horology*
March 1976, W. E. May and H. D. Howse, "How the Chronometer Went to
Sea," *Vistas in Astronomy* 20 (1976): 135−137, and E. G. R. Taylor, *Navigation
in the Days of Captain Cook*, National Maritime Museum Monograph no. 18
(London, 1974), all well annotated, can be usefully studied for the practical
aspects of navigation at that time.

44. D. H. Sadler, *Man is Not Lost* (H.M.S.O., London, 1978; first published 1968) is devoted to the history of the *Nautical Almanac*, 1767−1967,
and of lunar distances. It analyzes and reprints pages from *The Nautical
Almanac* for 1767—the first edition.

45. Richard Eden, *The Arte of Navigation*, trans. Martin Cortes (London,

1561), epistle, fol. *iii* verso; this was the first manual of navigation printed in English; see D. W. Waters, *The Art of Navigation in England in Elizabethan and early Stuart Times*, (London: Hollis and Carter, 1958; 2d ed., National Maritime Museum, Greenwich 1978), pp. 75–77 and 104.

46. J. Ramsden, *Description of an Engine for Dividing Mathematical Instruments* (London, 1777). He developed it between 1771 and 1773. The machine was capable of dividing both circular arcs and linear scales. It employed principles already published by the French Academician Duc de Chaulnes (1714–1769). It is illustrated and its principles explained in fig. 378 D. J. Price, "The Manufacture of Scientific Instruments from c. 1500 to c. 1700," in C. Singer et al., *A History of Technology*, Vol. III (Oxford: Clarendon Press, 1957). Price states: "it opened up a new era in precision workmanship and a new concept of instrumental accuracy. Thenceforward the technological problems of instrument making have been a determining factor in the progress of scientific research." It was as fundamental a mechanical advance in precision angular and linear measurement as Harrison's horological advances were in precision time measurement. Both were the fruits of the Longitude Act. Ramsden received an award of £615 from the Board of Longitude for his invention; see A. Stimson, "The Influence of the Royal Observatory at Greenwich upon the Design of Seventeenth- and Eighteenth-century Angle-Measuring Instruments at Sea," *Vistas in Astronomy* 20 (1976): 123–130 (p. 128).

47. The most complete and authoritative history of the observatory is *Greenwich Observatory: The Royal Observatory at Greenwich and Herstmonceux 1675–1975* (London: Taylor and Francis, 1974), vol. 1, *Origins and Early History (1675–1835)*, E. G. Forbes; vol. 2, *Recent History (1836–1975)*, A. J. Meadows; vol. 3, *The Buildings and Instruments*, D. Howse. D. Howse, *Francis Place and the Early History of the Greenwich Observatory* (New York: Science History Publications, 1975), reproduces the twelve etchings (exceedingly rare) of the Royal Observatory made shortly after its completion by Francis Place from drawings by Robert Thacker.

48. Edmund Spenser, *The Faerie Queene*, I, *ix*. 40.

VII

TORY–HIGH CHURCH OPPOSITION TO SCIENCE AND SCIENTISM IN THE EIGHTEENTH CENTURY: THE WORKS OF JOHN ARBUTHNOT, JONATHAN SWIFT, AND SAMUEL JOHNSON

Richard G. Olson

According to one recently developed perspective in the history of science, Newtonian natural philosophy—in contrast with Aristotelian, Epicurean atomist, Cartesian, and hermetic variants—and latitudinarian Anglicanism—as opposed to enthusiastic dissent, free-thinking deism, or high-flying Anglicanism—both achieved dominance in the early eighteenth century in England because they alone served the interests of trade and commerce which were being forwarded after the Glorious Revolution by the Whig political party. Furthermore, according to this point of view, the fortunes of Whig, Newtonian, and latitudinarian doctrines were intimately and self-consciously united as parts of a single ideological structure.

Margaret Jacob has summarized the role of scientific thought in this ideological complex as follows:

The ordered, providentially guided, mathematically regulated universe of Newton gave a model for a stable and prosperous polity ruled by the self interest of men. That was what Newton's universe meant to his friends and popularizers: it allowed them to imagine that nature was on their side. . . .

The Latitudinarians grafted the New Philosophy on to their social ideology, integrating both into English thought at precisely the time when modern and capitalistic forms of economic life and social relations were gaining ascendancy.[1]

I remain unconvinced that its ideological role is either necessary or sufficient to account for the acceptance and rise to dominance of Newtonian natural philosophy within the growing community of scientists of early eighteenth-century Britain. But it is undeniable that Broad Church Anglicanism, Newtonian science, and Whig political attitudes were mutually supportive and frequently expounded together by individuals at the beginning of the century. Political radicalism and enthusiastic religion were, moreover, often linked in contemporary minds with hermetic and Baconian science; and free thought, deism, and even atheism were closely tied to the materialistic natural philosophies of the Epicureans, Cartesians, and Hobbists. Given these widespread linkages, we might expect that Tory political opinions and High Church religious affiliations should have been linked with generally antiscientific sentiments, and that a special antagonism should have been directed at Newtonian science, since it was most directly associated with the Whig cause. This is particularly plausible because the Tories were fighting a rearguard action against values associated with rising commercial capitalism.

It will be my purpose to argue that there was indeed substantial antiscientific sentiment expressed by leading Tory–High Church intellectuals during the first two-thirds of the eighteenth century. Moreover, it will be my aims to (1) illuminate both the motives for and the character of that sentiment and (2) show why it was *not* principally anti-Newtonian, by considering the writings of three men, John Arbuthnot, Jonathan Swift, and Samuel Johnson.

One set of important and potentially troublesome issues—that of determining precisely the evolving meaning and significance of the terms *Tory, High Church, Whig,* and *latitudinarian* during the time under consideration—does not arise, because the three figures I am considering all identified themselves with, or willingly allowed themselves to be identified as, both Tories and High Church sympathizers. I cannot completely avoid a second problem of terminology, however—that associated with the use of *antiscientific*, a term none of these men used or would have recognized. In general, the term *antiscientific* will be applied to statements that disparage scientific theories, methods, activities, and attitudes, not because they fail to meet some technical criterion of adequacy—for example, that the theory fails to account for some particular phenomenon, that the method lacks precision, or that the activity has failed to generate the specific information its adherents claimed as its fruits—but rather because by holding the theories or using the methods, engaging in the activities or expressing the attitudes, men are led either directly or indirectly to bring harm to themselves or others. Only in rare cases (i.e., in connection with the military uses of what we would identify as mere technology but what many early eighteenth century figures saw as the products of scientific knowledge) was the perceived harm both direct and physical. Much more frequently harm was seen to derive from the pridefulness and moral insensitivity that seemed to accompany scientific theorizing, or from the attempt to extend scientific approaches to inappropriate domains—especially to religious and closely associated moral issues.

Between obviously technically motivated critiques of specific methods and theories and very broad moralistic attacks on virtually all forms of natural philosophizing, there is a huge range of instances in which what may seem to one interpreter to be fundamental attacks on science may look to another to be nothing but trivial and superficial responses to a lunatic fringe that never was honestly scientific in the first place. Most eighteenth-century criticisms of science lie in this ambiguous area. And to make matters worse, the satirical form in which many such criticisms were couched depended on absurd exaggeration for its rhetorical impact; so attacks on scientific attitudes and activities were almost always accompanied by obviously unwarrantable extensions of the ideas and activities of responsible figures. Given these circumstances, it is extremely easy to underestimate the seriousness and cogency of eighteenth-century antiscientific claims, either by viewing them as frivolous because they were not directed at real science, or by seeing them as unintelligent because they suggest a confusion of solid scientific substance with ridiculous extrapolation. For these reasons, Thomas Babington Macaulay, writing in the mid-nineteenth century, found it possible to write off virtually all the comments I am about to consider by saying:

It is true that the follies of some persons who, without any real aptitude for science, professed a passion for it, furnished matter of contemptuous mirth to a few malignant satirists who belonged to the previous generation, and were not disposed to unlearn the lore of their youth.[2]

Most subsequent interpreters have followed Macaulay's lead.

Without denying the substantial germ of truth in this, I wish to emphasize two fundamental facts. First, there is abundant evidence to show that assorted "virtuosi" and "projectors" saw themselves, and were seen by, seventeenth- and eighteenth-century figures as engaged in real natural philosophy or science. For us to define science in this period as what was done by Isaac Newton and a handful of others is to be horribly ahistorical. Men like John Wilkins and William Whiston, on the one hand, and Thomas Newcomen and Thomas Savery, on the other, were often lumped together by contemporaries as exemplifying what we would call a scientific approach to the world. Second, it has never been the case, nor is it ever likely to be the case, that public attitudes toward science derive from what a few professionals identify as the best science being done. For the purpose of understanding Tory and High Church attitudes toward science, the fundamental question is not whether men opposed what we *now* think was *good* science but whether they opposed what was then thought to be honestly scientific.

JOHN ARBUTHNOT

The leader of early eighteenth-century Tory–High Church commentators on science was undoubtedly John Arbuthnot. Born in 1667 to Alexander Arbuthnot, a Scottish Anglican clergyman who refused to conform to Presbyterianism when it was reinstituted in Scotland in 1689, John remained loyal to the Church of England; his brother Robert became a trusted agent of the pretender, moving to France with James II.

Arbuthnot attended Marischal College, Aberdeen, studying mathematics with David Gregory, and received an M.A. degree in 1685. A few years later his father was deprived of his living and John left for London, where he first came to public attention through his publication of a translation and expansion of Christiaan Huygens's *De Ratiociniis in Ludo Aleae* under the title *Of the Laws of Chance*. This work, in which Arbuthnot applied elementary probability theory to popular dice and card games, illustrates the moralizing tone of all his scientific writings: "My design in publishing this," he wrote, "was to make it of more general use and perhaps to persuade a raw squire to keep his money in his pocket; and if upon this account, I should incur the clamors of the sharpers, I do not much regard it since they are the sort of people the world is not bound to provide for."[3]

In 1697 he published his first criticism of a fellow scientist. Arbuthnot's *An Examination of Dr. Woodward's Account of the Deluge* was a technically motivated critique; but there were some religious comments and some passages of marvelously wry wit. Both of these characteristics deserve notice, for they were both to become more prominent in Arbuthnot's works as time went on. John Woodward's *An Essay Towards a Natural History of the Earth and Terrestrial Bodies, Especially Minerals; as Also of the Sea, Rivers, and Springs. With an Account of the Universal Deluge and of the Effects that It had Upon the Earth*, first summarized in the *Philosophical Transactions* for 1695, was one of the first putatively scientific attempts to explore the themes opened by Thomas Burnet's *Sacred Theory of the Earth* (1681–1690). Woodward, a pugnacious professor of physic at Gresham College, claimed to be a Baconian philosopher who built his work solely on the basis of careful observation. Indeed, the first section of his *Essay* was a powerful, though not very original, defense of the animal rather than mineral origin of marine fossils. But Woodward ranged far beyond his evidence to speculate about the origins of the world and the character of the Noachian Flood. Though very vague in detail, his theory was in many respects similar to that later attributed to Abraham Gottlob Werner by his Neptunist students.

Arbuthnot's attack on this work had a twofold strategy. First, he sought to show that Woodward's overly vague and general hypotheses were inconsistent with natural law when they were quantified and followed out in detail. Second, he sought to demonstrate that even if the theory could be reconciled with "the doctrine of secondary causes," it would nonetheless violate any intelligent and plausible reading of Scripture. Woodward's account was thus both bad science and bad theology; and it deserved censure on both accounts.

In order to show just how absurd Woodward could be, Arbuthnot calculated that in order to dissolve a bulk of earth equal to that supposed by Woodward, the volume of water would have to be that of a sphere extending at least 450 miles above the surface of the globe. But according to Moses—and to Woodward—the waters of the Flood rose only 15 cubits over the tops of the highest mountain. This amount of water, asserted Arbuthnot, "is too little to make the earth into an Electuary [broth], nay hardly into a pill."[4]

At several points in his critique Arbuthnot explicitly acknowledged that God, in his omnipotence, could have made things happen as Woodward supposed; but if He did so it was contrary to His ordinary method of operating through natural law. Among the most implausible aspects of Woodward's explanation was that the world should have been precipitated out of solution "into the same figure it had before the deluge, into equal Cavities and Eminences and alike scituated with those it had then."[5] This must have been the case or else Noah would surely have noted the great changes that took place; but Noah seemed to have found the mountains just where he left them. Arbuthnot concludes with two central thoughts: (1) that "the compilers of theories would have more regard to Moses' Relation, *which surpasses all the accounts of philosophers as much in Wisdom as it doth in Authority*,"[6] and (2) that "then only we may expect to succeed in the compiling of theories, when we build upon true and decisive observations; and survey the works of nature with the same Geometry (though in a more imperfect degree) by which the divine Architect put them together."[7] The second of these sentiments could have been expressed by virtually any latitudinarian natural theologian; but the first could only have been expressed by one whose primary commitment was as much to the scientific validity of revelation as to that of reason.

In the thirteen years following his criticisms of Woodward, Arbuthnot's published works developed the second line of argument to the near exclusion of the first. While Arbuthnot was developing his scientific reputation, however, he began another literary career that was to overshadow his work as a mathematical philosopher. As physician to Queen Anne, beginning in 1705, Arbuthnot developed close associations with major Tory political figures. By 1710 he had become an important advisor to the Harley-Oxford ministry, and in 1712 he wrote a series of political satires, the *John Bull* pamphlets, which introduced one of the great and lasting figures of British political life and which helped build sentiment against continued Continental war and in favor of the Peace of Utrecht. About the same time, his *Art of Political Lying* attacked the character of political discourse—that of contemporary Whigs in particular.

Though neither *John Bull* nor *Political Lying* addressed directly the character of contemporary science, in each of them Arbuthnot at least linked certain kinds of scientific and scientistic theorizing with negative sentiments. In *John Bull* he associated Jack (following Swift's *Tale of a Tub* in using this name to represent dissenting religion, especially Scottish Calvinism) with studies that "were generally bent towards exploded chimeras, the perpetuum mobile, the circular shot, philosopher's stone, silent gunpowder, making chains for fleas, and instruments to unravel cobwebs and split hairs."[8] Thus he linked religious dissent with the improving or projecting mentality of Baconian science and with the superstitions and quackery of alchemists. He simultaneously ridiculed the kind of science that had become associated with the terms *virtuoso* and *projector*. In *The Art of Political Lying* Arbuthnot beautifully parodied the Hobbist tendency to derive pernicious moral doctrines from scientific theories of human nature. According to the narrator of *Lying*, it is easy to understand why lying seems so much more natural and common among men than speaking the truth. This is so because the soul is constructed like a plano-cylindrical speculum, or looking glass: "The plane side represents objects just as they are; and the cylindrical side, by the rules of catoptrics, must needs represent true objects false, and false objects true: but the cylindrical side, being much the larger surface, takes in a greater compass of visual rays."[9] Since we know from elsewhere that Arbuthnot was especially impressed by the certitude of the science of geometrical optics, there can be no mistaking the thrust of the satirical passage: it is to challenge the unwarranted application of scientific analogies to moral arguments and the human soul.

As a Tory intellectual Arbuthnot met Jonathan Swift in 1711, and he joined with Swift, Matthew Prior, John Freind (another Tory Newtonian), and others in the Brothers Club of Tory wits. In 1713 a chance event occurred that greatly extended his literary influence and focused his satirical concerns on scientific topics. Alexander Pope approached Swift with a proposal that they and their mutual friends collaborate on a monthly burlesque of the follies of learning and criticism.[10] Though a variety of pressures kept Pope's original plan from developing in detail, Pope, John Gay, Thomas Parnell, Robert Harley, Swift, and Arbuthnot met together off and on throughout early 1714 and corresponded for over fifteen years to collaborate on the works of Martinus Scriblerus, which grew out of Pope's proposal. According to the modified plan, the Scriblerus project was to center on the fictional autobiography of the learned and universal—but misguided—genius Martinus Scriblerus. After the group had established Martin collectively through this device, each member would publish satires in his own domain of expertise in Martin's name. Seriously intended but particularly outrageous works written by others would also be claimed for Martin by his creators.

The Memoirs of the Extraordinary Life, Works, and Discoveries of Martinus Scriblerus finally appeared only in 1741, in the second volume of Pope's *Prose Works*, and only a few short essays were actually published in Martin's name; the death of Queen Anne and the consequent fall of the Harley ministry in late 1714 dispersed the group. The *Memoirs* are interesting in their own right, however, and the project almost certainly gave rise to at least two major works, Pope's *Dunciad* and Swift's *Gulliver's Travels*, both of which were initially projected as Martin's publications.

All of the authors involved acknowledged that Arbuthnot was primarily responsible for the content of the *Memoirs*.

When the group split up in June of 1714, Swift wrote to Arbuthnot: "To talk of Martin in any hands but yours is folly. You every day give better hints than all of us could do together in a twelve month; and to say the true, Pope, who first thought of the hint has no genius at it, in my mind. Gay is too young, Parnell has some ideas of it, but is idle; I could put together and Lard, and strike out well enough, but all that relates to the sciences must be from you."[11] Though Pope seems to have done the final editing rather than Swift, the latter's judgment about Arbuthnot's role was sound.

Because Arbuthnot was the principal author, the *Memoirs* are particularly rich in satire of contemporary science. Arbuthnot took aim at a broad spectrum of scientific follies, not excluding those of the mathematically oriented Newtonians and the political economists, with whom he had previously been associated. Thus, for example, he parodied William Whiston's and Humphrey Ditton's proposals for determining longitude by constructing a world-encircling system of fire ships that would send up flares exactly at noon each day.[12] And he chided the political economists for their tortuous and indirect attempts to measure simple and easily determined quantities; he suggested, for example, that the population of London be determined by measuring the amount of excrement collected.[13] Some of his targets, like Whiston's longitude project, were simply absurdly impractical schemes; and many of his individual complaints could have been made by latitudinarians against the strains of science linked to Hobbesian atheism or radical dissent; but on balance, Arbuthnot's criticisms were more broadly aimed at the excessive zeal and philosophic pride of *all* forms of contemporary science. He was deeply concerned that men, filled with a sense of their own knowledge, might think science sufficient for all human ends and negate any need for God and his revelations.

Like all literary parodies dealing even vaguely with scientific topics, the *Memoirs* attacked the lack of discrimination and sensitivity of the empiricist virtuosi. Consider just one of many possible examples. As Martin grew up with an indiscriminate love of learning and of curiosities, he was virtually immune from the usual passions of youth until he met "the most beautiful curiosity of nature, . . . two Bohemian sisters, whose common parts of generation had so closely allied them that Nature seemed to have conspired that their lives should run in an eternal parallel."[14] Martin was smitten, for, as Arbuthnot writes, "how much so ever our Martin was enamoured of her as a beautiful woman, he was infinitely more ravished with her as a charming monster. . . . [He was] unable to resist at once so pleasing a passion and so amiable a phenomenon."[15] Martin even wrote to his love: "Ages must be numbered, nay, perhaps some comet may vitrify this Globe upon which we tread, before we behold a Castor and Pollux resembling the beauteous Lindamira and Indamora. Nature forms her wonders for the wise, and such a master piece she could design for none but a philosopher."[16]

There was clearly something grotesque and perverted about a person whose priorities in love gave precedence to philosophical curiosity over the pleasures of normal human sexuality, and Arbuthnot's treatment of this theme allowed him to emphasize the traditional disturbing tendency of virtuosi to be fascinated by the abnormal and monstrous.

There was nothing peculiarly unique or conservative about Arbuthnot's criticisms of the virtuoso mentality; nor was there anything very unusual about his choosing materialist theories of the soul and mind for extensive satirical treatment. Beattie, his nineteenth-century biographer, claims that Arbuthnot did not see materialism as an enemy of religion, "as he might have done later in the century."[17] It is very clear, however, that antimaterialism was an essential part of both latitudinarian and High Church attacks on atheism because of Hobbes's supposed linking of materialism and atheism and because of the long-standing ties between ancient materialism and atheism. What distinguished Arbuthnot's critical attitude toward materialism from that of his latitudinarian colleagues was the depth of his skepticism regarding the capacity of human reason. In his extreme skepticism, he was centrally in the High Church camp. Indeed, one of his targets of ridicule was the latitudinarian and Newtonian Samuel Clarke, whose self-assured disproofs of the materiality of the human soul drew almost as much fire from Arbuthnot as the counterproofs of Anthony Collins.

Arbuthnot's own belief was precisely that held by virtually

all High Churchmen and articulated most concisely by his friend Swift: "The manner whereby the soul and body are united, and how they are distinguished is wholly unaccountable to us. We see but one part, and yet we *know* we consist of two; and this is a *mystery* we cannot comprehend any more than that of the *Trinity*."[18] The linkage of doctrines of the soul and of the Trinity is significant; on both issues the latitudinarians and High Churchmen split over the relative importance of reason and revelation. The latitudinarians' appeal to reason both ensured them of the immaterial soul's existence and led many of them, including Newton, Whiston, and Tillotson, to doubt the doctrine of the Trinity. High Churchmen, by contrast, tended to consider both questions to be hopelessly beyond the reach of human reason. Arbuthnot explored this issue most extensively and most clearly in his serious poem *Know Thyself*, which was probably written sometime around 1715 but appeared only in 1734. Here he reviewed again the central themes of the Collins-Clarke debates, but with no pretense at humor, concluding once more that neither argument was adequate or convincing regarding the nature of the soul. In *Know Thyself* Arbuthnot went on, however, to offer his own answer to the question of soul-body interaction, which lay in the revealed word of God alone.

> Almighty power, by whose wise command,
> Helpless, forlorn, uncertain, here I stand;
> Take this faint glimmering of thyself away,
> Or break into my soul with perfect day!
> This said, expanded lay the sacred text,
> The balm, the light, the guide of souls perplexed . . .
> I grope and guess no more, but see my way;
> Thou cleardst the secret of my high decent,
> And told me what those mystic tokens meant;
> Marks of my birth, which I had worn in vain,
> Too hard for worldly sages to explain.

Lest anyone miss the message, Arbuthnot concluded with one final reminder to get priorities straight and to give up the philosophical pride that stands in the way of accepting the word of God.

> In vain thou hop'st for bliss on this poor clod,
> Return, and seek thy father and thy God;

> Yet think not to regain thy native sky,
> Born on the Wings of vain philosophy. . . .
> Let humble thoughts thy wary footsteps guide.
> Regain by meekness what you lost by Pride.[19]

Arbuthnot was probably more prolific in suggesting targets for ridicule and correction than any of his contemporaries; he recognized the vulnerable spots in the scientific enterprise both because of his intellect and because of his intimate knowledge of the scientists of his time. Nevertheless, he seldom developed really brilliant and lasting images. As the criticisms of the moral and religious implications of scientific activity passed to others, however, they were handled not simply with greater literary skill but often with a more strident and desperate tone. What began for Arbuthnot as one scientist's concern for the unintended errors and excesses of his colleagues, in the hands of Jonathan Swift turned into a full-blown antagonism to the scientific enterprise—an antagonism that was particularly pointed in connection with the kind of quantifying mentality that Arbuthnot had seen as an unassailably favorable feature of scientific thought. It is little wonder that Arbuthnot should have been distressed at how his "hints" had been developed and directed in the voyage to Laputa, or that he wrote to Swift praising *Gulliver's Travels* in general but grumbling that "the part of the projectors is least brilliant."[20]

JONATHAN SWIFT

Like Arbuthnot, Jonathan Swift was born in 1667 into the family of a conservative Anglican clergyman who lost his living rather than compromise his principles. The head of the family, Jonathan's grandfather, gathered his sons together and fled to Ireland during the civil wars. The crucial factor, however, in determining Swift's political and religious attitudes as well as his stance toward scientific pursuits was his stay with Sir William Temple. The relationship between Swift and Temple—which Swift broke off for two years between 1694 and 1696—was a stormy one; but between 1689, when he entered Temple's service as secretary, and 1699, when Temple died, Swift spent most of his time with Temple and accepted most of Temple's basic attitudes as his own.

Sir William Temple had been a Whig with powerful political connections, but by the 1690s his basic sentiments were those adopted by the moderates in the Tory party. In particular, Temple's urbanity was overlaid with a deep and explicit appreciation for the rural life. Furthermore, and even more importantly, Temple's religion was High Church Anglicanism of a particularly shrewd antiphilosophical and antidissenter variety. According to his sister it "was that of the Church of England which he was born and bred in, and he thought nobody ought to change, since it must require more time and pains than one's life can furnish to make a true judgement of that [i.e., nonconformist opinion and latitudinarian compromise] which interest and folly were commonly motives to."[21]

There is no reason to question Temple's religious sincerity; but he clearly felt that true religion was a matter of faith and/or habit which could not be touched by learning, force, or reason. Thus, he argued "it is too magistral . . . in matters of religion to impose anything on a man's *belief* . . . I can, if that will serve the turn, put in a glass eye, put on a false nose, take up a seeming belief, but not a real one. To say *you* can believe it signifies nothing to me. So perhaps you can see a mile off, when I am purblind and cannot see an ell."[22]

If religious belief came close to being a matter of individual conscience for Temple, however, as it was for some dissenting sects, institutional religion and religious behavior was both extremely important and a public rather than a private matter. Open dissent was inexcusable as subversive of the stability of society. Thus, formal conformity to Anglicanism was not just a reasonable but a necessary prerequisite for the British citizen, and even more so for the public officeholder.

From the first of his youthful poems written in Temple's household to the last of his polemical tracts, Temple's protégé Swift held to this basic attitude toward religious belief and behavior, which linked an absolutely nonromantic but transcendent faith with a rigid civic conformity. In his *Thoughts on Religion* Swift wrote:

To say a man is bound to believe, is neither truth nor sense.

You may force men, by interest or punishment, to say or swear they believe: you can go no further.

Every man, as a member of the commonwealth, ought to be content with the possession of his own opinion in private, without perplexing his neighbor or disturbing the public.

Violent zeal for truth hath an hundred to one odds to be either petulancy, ambition, or pride.[23]

Just as Swift followed Temple in focusing on the externalities of religion because belief and true piety seemed to be beyond the reach of force and argument, so he followed Temple in criticizing modern learning, especially scientific learning, because he felt that the most important issues—those of true piety and virtue—were certainly beyond the range of human investigations of nature. In 1690 Temple wrote his *Essay on Ancient and Modern Learning*, in which he downplayed the recent achievements of natural philosophy primarily on the grounds that they simply diverted men from the more critical tasks of living virtuously and piously. As he said in one of his reviews of the essays of his chief critic, William Wotton, F.R.S., "human learning seems to have very little to do with true Divinity."[24]

Moreover, for Temple, as for Arbuthnot, modern scientific knowledge, while incapable of providing important moral and religious guidance, was both capable and demonstratively effective in producing a vain, presumptive, and destructive sense of pride in intellectual accomplishment:

Our [modern scientific] learning leads us to presumption, and vain ostentation of the little we have learned, and makes us think we do, or shall, know, not only all natural, but even what we call supernatural things; all in the heavens, as well as upon the earth; more than all mortal men have known before our age; and shall know in time as much as angels.[25]

No doubt this kind of argument had a special appeal to a young man who had performed dismally in his academic work in natural philosophy.

Whatever the reasons, the impact on Swift's ideas of Temple's attitudes toward scientific learning seems to have been immediate, intense, and lasting. The year after Temple wrote his *Essay*, Swift was already parroting its sentiments in his poetry:

Look where exalted Virtue and Religion set
Enthroned with heavenly Wit,

Look where you see
The greatest scorn of learned vanity . . .[26]

His opposition to philosophical speculation, or "unrevealed philosophy"—whether ancient or modern—on the grounds of its irrelevance for and corruption of true religion, remained a main theme of his sermons in later life. In "On the Wisdom of this World," for example, he emphasizes Christ's antagonism toward the Pharisees and Sadducees, "who followed the doctrines of Plato and Epicurus," and he cites and analyzes Paul's letters to the Colossians and to Timothy, which counsel us to "avoid profane and vain babblings, and oppositions of science" because the secular scientists err concerning the faith and because disputes among philosophers lead to disputes, dissension, uncertainty, and skepticism even regarding sacred issues.[27]

In some sense Swift's lifelong concern with science was only occasional and incidental and his scientific knowledge superficial. In spite of the demonstration by professors Nicolson and Mohler that most of the themes in book three of *Gulliver's Travels* derive directly from ideas presented in the *Philosophical Transactions*,[28] and in spite of the more recent claims by Colin Kiernan that Swift's "scientific synthesis found a middle way between the extremes of Newtonian and Paracelsian science,"[29] there seems to me no reason to believe that the distinctions between Newtonian, Cartesian, Paracelsian, and Epicurean science, which were central to the latitudinarians as well as to both Temple and Arbuthnot, penetrated Swift's awareness at all. As far as he was concerned, all scientific or philosophical systems, including that of Newtonian gravitational attraction, were equally based on conjecture. Each was a new fashion that would have its vogue and pass into oblivion.[30]

Swift was above all a divine, a moralist, and a serious political pamphleteer; so it was only in a few works immediately influenced by Temple during the last decade of the seventeenth century and in *Gulliver's Travels*—which was an outgrowth of the Scriblerian project and of Arbuthnot's scientific concerns and "hints"—that Swift focused on scientific themes at all. Nonetheless, *A Tale of a Tub* and *Gulliver* were Swift's most widely read and lasting works; and they contain classic statements of conservative antiscientific sentiment.

Substantial portions of *A Tale of a Tub* were Swift's defenses of Temple against the criticisms of the Royal Society's semi-official spokesman, William Wotton. In the second paragraph of the *Tale* Swift lets fly at the speculations of natural philosophers by referring to the first great satire of cosmological speculation, Aristophanes' *Clouds*.[31] In order to stand out above the masses, "the philosopher's way, in all ages has been by erecting certain edifices in the air: but whatever practice and reputation these kind of structures have formerly possessed, or may continue in, not excepting even that of Socrates, when he was suspended in a basket to help contemplation, . . . they seem to labour under two inconveniences. First, that the foundations being laid too high, they have often been out of sight, and even out of hearing. Secondly, that the materials being very transitory, have suffered much. . . . "[32] After this crack at the airy content of philosophical speculations, Swift attacks the parentage and seriousness of the Royal Society of London by satirizing it as a "junior society" spawned by Grub Street literary hacks and dedicated merely to competing with Grub Street for output of worthless literary productions.[33]

Swift depicts the uselessness of natural philosophizing both through mock praise of "the great usefulness of [William Wotton's] sublime discoveries upon the subjects of flies and spittle,"[34] and by his straight-faced complaint that "whatsoever philosopher or projector can find out an art to sodder and patch up the flaws and imperfections of nature, will deserve much better of mankind, and teach us much more useful science than that so much in present esteem of widening and exposing them (like him who held anatomy to be the ultimate end of physic)."[35] Finally, the hubris of the scientific theorists is portrayed in the magnificent digression on madness, in which Swift claims that if Epicurus, Apollonius, Lucretius, Paracelsus, Descartes, and others like them could be caught and separated from their followers, they would

incur manifest danger of phlebotomy, and whips and chains, and dark chambers, and straw, for what man, in the natural state or course of thinking, did ever conceive it in his power to reduce the notions of all mankind exactly to the same length, and breadth, and height of his own? Yet this is the first humble and civil design of all innovators in the empire of reason.[36]

Normal human beings are happy to go along with the mental patterns common in their society, rather than seeking to innovate, for the more they learn, the more they recognize that every new system of thought is idiosyncratic, faulted by the limitations of individual vision. But, says Swift:

when a man's fancy gets astride on his reason; imagination is at cuffs with the senses, and common understanding, as well as common sense, is kicked out of doors, the first proselyte is himself; and when that is once accomplished, the difficulty is not so great in bringing over others . . . because imagination can build noble scenes and produce more wonderful revolutions than fortune or nature will be at expense to furnish.[37]

At this point Swift exposes what may be his most fundamental reason for objecting to scientific system building. Precisely the same "mad" impulse that leads to scientific speculation leads also to religious deviance, or "enthusiasm," and to political revolution. There is a subtle difference indeed, he writes, between "Alexander the Great, Jack of Leyden, and Monsieur DesCartes";[38] whether the madman thinks of "subduing multitudes to his own power, his reasons, or his visions," is almost accidental. To encourage this kind of madness in one domain is to seriously risk its extension to the others. Thus, scientific pride will certainly be transformed into religious nonconformity and political revolution. In one of the most bitterly ironic statements of the *Tale*, Swift hints that, in fact, that has already happened in connection with growing free thought. Without the help of this pride and vanity, he writes, "the world would not only be deprived of those two great blessings, *conquests* and *systems*, but even all mankind would be reduced to the same belief in things invisible." That is, without the impulse that seems to be encouraged in speculative science we might have a unified and spiritual Christian religion without materialist challenges.

The final science-related issue broached by Swift in his earliest works was the seventeenth-century interest in natural theology, with its central focus on the wise and intentional design of the universe and its frequent appeal to microcosm-macrocosm analogies. In *A Tale of a Tub* Swift describes a sect that worships an idol in the form of a tailor and holds that the universe is a marvelous, large suit of clothes created by him.

Look on this globe of earth, you will find it to be a very complete and fashionable dress. What is that which some call land but a fine coat faced with green. . . . Proceed to the particular works of the creation . . . observe how sparkish a periwig adorns the head of the beech, and what a fine doublet of white satin is worn by the birch. To conclude from all, what is man but a micro-coat, or rather a complete suit of clothes with all its trimmings.[39]

Virtually no element of contemporary natural theology is absent from this detailed, good-humored, and at times ribald parody. John Ray's emphasis on organic imagery in his *Wisdom of God Manifested in the Creation* (1691) is there in references to the "vegetable beaux"; the Cambridge Platonists' emphasis on plastic nature acting on a minor level to continue creation is there in "journeyman Nature"[40] trimming up the creation; the artisan God of such mechanical philosophers as Boyle is the tailor himself; and the latitudinarian tendency to identify religious morality with self-interest is parodied in a marvelous passage on the perfect design of human mental characteristics or faculties:

to instance no more; is not religion a cloak; honesty a pair of shoes worn out in the dirt, self-love a surtout; vanity a shirt; and conscience a pair of breeches; which, though a cover for lewdness as well as nastiness, is easily slipt down for the service of both.[41]

Swift returned to each of the scientific themes in *A Tale of a Tub* a quarter of a century later in *Gulliver's Travels*; but with the passage of time his youthful wit had turned increasingly bitter. In the public sphere, his preferment in the church had been blocked—in part because of Anne's response to the irreverence of the earlier work—his Tory friends had been driven from office, and he increasingly identified with Ireland, which received nothing but exploitative treatment from its English master. In addition, his personal life was, to say the least, frustrating, and he was increasingly fearful that he might be going mad as a result of what we now think was a severe inner ear problem. Moreover, though the optimism of seventeenth-century projectors might have been barely palatable in the 1690s, it was much less so in the wake of the South Sea Bubble in the 1720s.[42]

When Swift returned under Arbuthnot's urgings to the scientific issues he had playfully dealt with before, he saw them magnified in importance, and with intensified anger. Earlier, Swift had merely given us Socrates suspended in a basket. Now he created a whole race of absentminded mathematical speculators suspended in a flying Island.[43] In place of the depiction of the Royal Society of London as a Grub Street offshoot, he offered the Academy of Lagado, which had been instituted by a group of Balnibarbians who had visited Laputa and come back "full of volatile spirits acquired in that airy region."[44] To Swift, by the 1720s the simple uselessness of natural philosophy had become a positive impediment to practical accomplishments. Thus, on the flying island, "Their houses are very ill built, the Walls bevil, without one right angle in any apartment; and this defect ariseth from the contempt they bear for practical geometry, which they despise as vulgar and mechanikal."[45]

In his visit to Brobdingnag, Gulliver offers Swift the opportunity to go even farther with his attack on the inutility of scientific and technical progress. Gulliver offers to serve the Brobdingnagian king by letting him in on the most advanced European military techniques. When he pulls out the principal standby of modern apologists for progress through innovation—guns and gunpowder—praising their destructive capabilities, the admirable king is appalled.

He was amazed how so impotent and groveling an insect as I (these were his expressions) could entertain such inhuman ideas, and in so familiar a manner as to appear wholly unmoved at all the scenes of Blood and Desolation, which I had painted as the common Effects of those destructive Machines; whereof he said some evil genius, Enemy to Mankind, must have been the first contriver.[46]

The products of modern learning are thus recognized as malevolent rather than benevolent or simply ridiculous. And the Brobdingnagian inability to see with the mad perspective of contemporary Europeans has, Gulliver writes, "risen from their Ignorance; by not having hitherto reduced *Politicks* into a *Science*, as the more acute wits of Europe have done."[47] Even political and moral judgments are corrupted by being made scientific—by being made in the name of *interest* rather than of Christian charity. Thus we have returned, in a new and more trenchant and direct way, to the linkage between scientific speculation and the undermining of religious and political life which had been suggested in *A Tale of A Tub*.

The issue of natural theology also remains, and it gives rise to several very different musings—both clever and bitter—throughout the *Travels*. In Lilliput the watch- or clockmaker imagery that dominated one stream of natural theology from Boyle through Paley is wryly twitted when the Lilliputians mistakenly interpret Gulliver's watch as the God he worships.[48] It is in Brobdingnag, however, that the fundamental issues of natural theology are faced most directly.

One of the most frequently voiced appeals of late seventeenth- and early eighteenth-century natural theologians was to the precision of design recognizable through microscopic observations. The paradigmatic statement of this appeal was certainly John Ray's presentation in *The Wisdom of God Manifested in The Works of Creation*.

Whatever is natural, beheld through [the microscope] appears exquisitely form'd, and adorn'd with all imaginable Elegancy and Beauty. There are such inimitable gildings in the smallest Seeds of Plants, but especially in the parts of Animals, in the Head or Eye of a small Fly; such Accuracy, Order and Symmetry in the frame of the most minute creatures, a *Louse*, for example, or a Mite, as no Man were able to conceive without seeing of them. Whereas the most curious works of Art, the sharpest and finest Needle, doth appear as a blunt rough bar of iron, coming from the furnace or the forge: the most accurate Engravings or Embossments seem such rude, bungling, and deformed Work, as if they had been done with a Mattox or Trowel; so vast a difference is there betwixt the skill of Nature, and the Rudeness and Imperfection of Art.[49]

Against this vision of natural perfection Swift set Gulliver's views of the giant Brobdingnagians and their creatures.

But the most hateful sight of all was the Lice crawling on their clothes: I could see distinctly, the Limbs of these Vermin with my naked Eye, much better than those of a *European* Louse through a Microscope: and their Snouts with which they rooted like Swine. They were the first I had ever beheld; and I should have been curious enough to dissect one of them, if I had proper Instruments (which I unluckily left behind me in the Ship) although indeed the Sight was so nauseous, that it perfectly turned my Stomach.[50]

Seen from this Swiftian perspective, nature is gross and horrible, "bungling and deformed." It suffers by comparison with any reasonable example of artisanship. The last thing that a louse suggests is order and symmetry, and the argument from design as presented by most natural theologians becomes just one more product of a limited human point of view. Certainly it could not be trusted as a cornerstone of Christian belief. To suggest that it should—as all scientistic latitudinarians were inclined to do—was more likely to undermine than to enhance Christian faith.

In some ways Jonathan Swift comes dangerously close to being Margaret Jacob's typical High Churchman, with an "ignorant and obscurantist opposition to everything new and modern." Nevertheless—though his antagonism to scientific speculation may have failed to discriminate among what others considered radically different kinds of science; and though his disappointment over the failure of applied science to deliver on some of its promises blinded him to the potential utility of the improving, or projecting, mentality—Swift's insights into the importance of tradition and the extreme limitations of human knowledge as derived from experience and shaped by philosophical systems, were very shrewd. They allowed him to recognize, as Arbuthnot could not, the severe dangers of allowing religious and political life to become scientized. Arbuthnot had sought to apply scientific arguments to religious and political topics and simply warned against the scientist's kind of pride in system that at times allows him to presume that mere human knowledge is adequate for a Christian life. Swift, by contrast, saw science as being more dangerous the more its techniques are extended to political and religious concerns, because he saw excessive pride as endemic, rather than accidental, to scientific speculation, and because, like Pascal and Rousseau (in *The Discourse on Arts and Sciences*), he saw religion and morality as simply part of a different order of existence, one that is forever untouchable by the reason of natural science.

SAMUEL JOHNSON

Born in 1709 to a Midlands bookseller and his dissenting wife, Samuel Johnson had no clearcut or obvious basis in family tradition for an orientation toward Tory politics and

High Church Anglicanism; and it is unclear why he should have adopted either position during a period in which most intellectuals were becoming Whigs and latitudinarians. Donald Greene, author of the most extensive recent analysis of Johnson's political thought, suggests that he took to Toryism primarily to show his independence of fashion.[51] The same argument may also hold for his religious opinions; but it is clear that his commitment to High Church Anglicanism became much more intense and conventional than his on-again, off-again Toryism. A commitment to a form of Augustinian Christianity based strictly on Scripture and with supreme emphasis on Christian love, charity, and humility, coupled with a supreme detestation of human pride as the major source of all human misery, was the dominant underlying theme in all his most significant moral writings.

With regard to scientific knowledge Johnson seems to have adopted many of John Arbuthnot's attitudes—his enthusiasm for morally guided applications of science, for a Newtonian emphasis on secondary causes, and for quantitative precision, as well as his fear that scientific studies could lead to false pride and insensitivity to human needs. At the same time, Johnson shared with Swift an intense fear that a kind of madness might be the unavoidable concomitant of a full commitment to the scientific enterprise; moreover, he shared with him a strong awareness of the distinction between moral wisdom and natural knowledge. For this reason, Johnson, like Swift, was more often critical than supportive of latitudinarian natural theology.

Far more even than Arbuthnot, Johnson was a believer in intellectual and material progress. He clearly saw scientific knowledge as the central feature of that twofold advance. And he had little patience with the Swiftian ridicule of all projectors. A passage from his life of Sir Francis Drake, published in *Gentleman's Magazine* in 1740, illustrates this theme very well.

There are some men, of narrow views and grovelling conceptions, who, without the instigation of personal malice treat every new attempt as wild and chimerical, and look upon every endeavour to depart from the beaten track as the rash effort of a warm imagination . . . These men value themselves . . . upon inventing arguments against the success of any new undertaking, and where argument cannot be found, upon treating it with con-

tempt and ridicule. Such have been the most formidable enemies of the
great benefactors to mankind.[52]

We might well want to ask whether all kinds of scientific
learning seemed equally worthwhile to Johnson, just as they
seemed equally dangerous to Swift. On this point Johnson
leaves us in no doubt that he stands firmly with Arbuthnot in
the Newtonian tradition that focused on secondary causes and
mathematical argumentation. In his life of Boerhaave, the
great Dutch expositor of Newtonian methodology, Johnson is
unstinting in his praise of Boerhaave's attacks on scholastic,
Cartesian, and hermetic science.

The emptiness and uncertainty of all those systems, whether venerable for
their antiquity, or agreeable for their novelty. He has evidently shown; and
not only declared but proved, that we are entirely ignorant of the *principles* of
things, and that all the knowledge we have is of such qualities alone as are
discoverable by experience, or such as may be deduced from them by
mathematical demonstration.[53]

One would be hard pressed to discover a more complete
commitment to Newtonian methodology on the part of any of
Margaret Jacob's latitudinarian Newtonians. Furthermore,
Johnson retained a special concern with quantification
throughout his life. Like Arbuthnot, he saw quantitative precision as a great corrective to speculative and superstitious
error: thus, he wrote to Mrs. Thrale that "a thousand stories
which the ignorant tell, and believe, die away at once, when the
computist takes them in his gripe . . . never think you have
arithmetic enough,"[54] and he engaged in numerous Petty-
esque demographic speculations in his *Journey to the Hebrides*.

Everything I have mentioned thus far would make Johnson
one of the great apologists for Baconian and Newtonian science. On balance, however, I think he was actually the greatest
eighteenth-century critic of the scientific enterprise as he
understood it. To understand why, we must consider Johnson's sense of priorities and his sense of how certain antisocial
tendencies associated with scientific learning could get out of
control and turn destructive.

The starting point for this consideration lies in Johnson's
attitude toward natural theology and his perception of the
relationship of natural science to religion and morality.

Though these problems concerned him throughout his adult life, the most extended and articulate statement of his position appeared only in 1779 in his *Life of Milton*. In the midst of his discussion of the school Milton ran as a young man, Johnson introduces what seems at first to be a gratuitous attack on Milton's decision to include a certain amount of natural philosophy in the curriculum.

The truth is that the knowledge of external nature, and the sciences which that knowledge requires or includes, are not the great or the frequent business of the human mind. Whether we provide for action or conversation, whether we wish to be useful or pleasing, the first requisite is the religious and moral knowledge of right and wrong. . . . We are perpetually moralists, but we are geometricians only by chance. Our intercourse with intellectual nature; our speculations upon matter are voluntary, and at leisure . . .

Let me not be censored for this digression as pedantick or paradoxical: For if I have Milton against me, I have Socrates on my side. It was his labour to turn Philosophy from the study of nature to speculations upon life; but the innovators whom I oppose are turning off attention from life to nature. They seem to think that we are placed here to watch the growth of plants, or the motions of the stars. Socrates was rather of the opinion, that what we had to learn was, how to do good, and avoid evil.[55]

The most central feature of this statement is not explicit: it is the implicit assumption that natural philosophy has no direct or fundamental bearing on morality and religion. To hold this position after reading the major works of Boyle, Wilkins, Bentley, Tillotson, John Ray, and others, is to take a self-conscious position in opposition to latitudinarian natural theologizing. It is to directly renounce John Ray's claim that for a free man there is "no occupation more worthly and delightful than to contemplate the beauteous works of nature and honour the infinite wisdom and goodness of God"[56] and to take a position more consonant with that of the Evangelical clergyman William Romaine, whose works Hill Boothly had recommended to Johnson. Romaine argued that "the mighty volumes of visible nature, and the everlasting tables of right reason" were completely useless in the Christian search for salvation.[57] While Johnson detested the enthusiastic tone of Romaine's argument, he clearly accepted the separation between natural knowledge and religious wisdom which it implied.

That this was a long-standing sentiment with Johnson is suggested once more in his special praise of Boerhaave for refraining from natural theologizing. Boerhaave, Johnson tells us, was revolted when he "found the holy writers interpreted according to the notions of philosophers, and the chimeras of metaphysicians adopted as articles of faith."[58] Speaking of Boerhaave's own religious attitudes, he says:

A strict obedience to the doctrine, and a diligent imitation of the example of our blessed savior, he often declared to be the foundation of true tranquility . . . He worshiped God as he is in himself, *without attempting to inquire into his nature* . . . There he stopped lest, by indulging his own ideas, he should form a Deity from his own imagination and *sin* by falling down before him. To the will of God he paid absolute submission, *without endeavouring to discover the reason of his determinations*, and this he accounted the first and most inviolable duty of a Christian.

This is a complete projection of Johnson's own religious attitudes upon the one man about whom he was able to be completely uncritical.[59]

If Johnson frequently expressed his unease with natural theologizing, all of his other comments pale in comparison with his violent opposition to Soame Jenyns's extension of natural theology in *A Free Enqiury into the Nature and Origin of Evil*, published in 1757. Jenyns, who seems to have been a moderate Anglican in religion, sought with little originality to explain the nature of apparent evil in terms of the so-called great chain of being and on the assumption that the "beauty and happiness of the whole depend altogether on the just inferiority of the parts." Johnson points out that although innumerable naturalists seem to assert that the "doctrine of the regular subordination of beings, the scale of nature, and the chain of nature" flows from their observations of the world, his own considerations of this claim always leave him uncertain and doubtful. Furthermore, he says that, by implying that the nature of matter forces God into choosing the lesser of evils, Jenyns is guilty of placing "dogmatical limitations of omnipotence on God's power." Jenyns would do better to examine his own limitations instead of God's.[60]

These comments are just the prelude, however, to Johnson's attack on Jenyns's blasé argument that, just as human beings hunt animals for pleasure, there are invisible higher beings who "may deceive, torment, or destroy us for their own pleasure or utility."[61] This is an appalling thought that Johnson attacks by satirically extending it in a positively Swiftian manner.

As we drown whelps and kittens, they amuse themselves now and then with sinking a ship, and stand around the fields of Blenheim, or the walls of Prague, as we encircle a cockpit. . . . Some of them, perhaps are virtuosi, and delight in the operations of an asthma, as a human philosopher in the effects of the air pumps. To swell a man with a tympany is as good sport as to blow a frog.[62]

Only in one other piece of writing, and in connection with the same presumption that man might penetrate God's reasons for evil by following natural analogies, does Johnson's prose become this bitter. In the original *Idler*, 22 (suppressed when the essays appeared as a book), a family of vultures sits in wait for the carnage of battle to finish so they can swoop down to feed on the decaying human flesh. The mother explains to her children that nature has devoted men to the uses of vultures and has marvelously arranged a strange ferocity in them so that they will kill one another without eating their prey. When the children ask why men bother to kill at all if they aren't going to eat those they kill, the mother responds by presenting the pseudo-Cartesian theory that men are really just nonliving automata "created by a beneficent nature for the welfare of vultures."[63]

Johnson's revulsion toward such theodicies as that of Jenyns clearly grew out of his almost Franciscan sense of God's love and concern for all his creatures—a love and concern whose emulation was the height of Christian living, and one that was mocked by the presumption that God "designed" cruelty and viciousness into the world.

His inability to believe that God could be, or that man should be, insensitive to pain and suffering of any kind provides the key to understanding Johnson's most severe criticisms of natural philosophy, as distinguished from its scientistic extension into natural theology. For, Johnson seems to believe that commitment to the search for knowledge may well produce insensitivity to both human and other problems. To some extent this criticism is linked to the long tradition of criticism of the virtuoso's bizarre set of values; but Johnson shifts its focus in a new direction that presages Theodore Roszak's claim that the coldness of the "objective consciousness" is as much a product of, as a prerequisite for, scientific activity. This theme is most fully expressed by Johnson in *Idler*, 17, which is an attack on vivisection.

Among the inferior professors of medical knowledge is a race of wretches, whose lives are only varied by varieties of cruelty; . . . What is alleged in defence of these hateful practices, everyone knows; but the truth is, that by knives, fire, and poison, knowledge is not always sought, and is very seldom attained. . . . he that burned an animal with irons yesterday, will be more willing to amuse himself with burning another tomorrow. And if the knowledge of physiology has been *somewhat increased, he surely buys knowledge dear, who learns the* use of the lacteals at the expense of his humanity. It is time that universal resentment should arise against these horrid operations, *which tend to harden the heart, extinguish those sensations which give man confidence in man, and make the physician more dreadful than the gout or the stone.*[64]

According to Johnson no other scientific activity matches vivisection in numbing man's natural and laudable feelings of concern and compassion for his fellow creatures and his fellow man; but every tendency to emphasize *nature* over *life* pushes in that direction.

The result of emphasis on nature to the exclusion of human concerns is portrayed in *Rambler*, 24, in the verbal picture of Geldius, the intense student of nature who reacts to news of his brother's death at sea with a question about weather conditions and who responds to a fire in a nearby town with a lecture on combustion.

Thus lives this great philosopher, insensible to every spectacle of distress, and unmoved by the loudest call of social nature, for want of considering that men are designed for the succor and comfort of each other; that though there are hours which may be laudably spent upon knowledge not immediately useful, yet the first attention is due to practical virtue; and that he may be justly driven out from the commerce of mankind, who has so far abstracted himself from the species, as to partake neither of the joys nor griefs of others, but neglects the endearments of his wife, and the caresses of his children, to count the drops of rain, note the changes of the wind, and calculate the eclipses of the moons of Jupiter.[65]

Only as long as natural knowledge is subordinated to moral and human ends—that is, as long as science is done in the service of Christian charity—can it be valued; otherwise it is vain at best.

Yet the morally crippled scientist Geldius seems to me only the second most tragic figure in all of Johnson's writings; the "mad astronomer" in *Rasselas* is even more so. His case demands special consideration because with it Johnson returns to Swift's linkage of madness with science and suggests that, even when pursued in charity, scientific activities may—perhaps must—lead to debilitating hubris.

Prior to 1759 all of Johnson's comments on scientific activity were consistent with the Baconian presumption that the scientific search for power for the amelioration of the human condition, properly subordinated to Christian charity, was among the most admirable of human activities. That scientists should occasionally reach farther than they could grasp was seen as a mildly unfortunate but probably necessary condition for long-term and desirable progress. In the mad astronomer of *Rasselas*, however, we are presented with a man who, though he lacks no human virtue, still cannot avoid a kind of insanity associated with scientific activity. Rasselas's teacher, Imlac, introduces the astronomer with unstinting praise:

I have just left the observatory of one of the most learned astronomers in the world, who has spent forty years in unwearied attention to the motions and appearances of the celestial bodies and has drawn out his soul in endless calculations . . .

His integrity and benevolence are equal to his learning. His deepest researches and most favorite studies are willingly interrupted for any opportunity of doing good by his council or his riches . . . "though I exclude idleness, . . . I will never," says he, "bar my doors against charity. To man is *permitted* the contemplation of the skies but the practice of virtue is commanded."[66]

Though the astronomer has avoided all the obvious pitfalls, he is truly mad: he tragically believes he is responsible for "the regulation of the weather and the distribution of the seasons."

The sun has listened to my dictates, and passed from tropic to tropic by my direction; the clouds, at my call, have poured their waters, and the Nile has overflowed at my command; I have restrained the rage of the dog star, and mitigated the fevours of the crab.[67]

Moreover, the problem is not that he seeks any glory as a result of his presumed power—on the contrary, he feels it is a great burden.

The astronomer's mind has snapped not because of the relentless and pointless search for pure knowledge but because he wants so deeply to use his knowledge to help mankind:

". . . my daily observations of the changes of the sky led me to consider, whether, if I had the power of the seasons I could confer greater plenty upon the inhabitants of the earth. . . . As I was looking on the fields withering with heat I felt in my mind a sudden wish that I could send rain . . . I commanded the rain to fall, and . . . I found that the clouds had listened to my lips. . . ."[68]

Just as the physician Aesclepias was punished by Zeus for seeking to become a deity himself by raising the dead in opposition to nature, so the astronomer has been driven to madness by his well-intended but overweening desire to be a benefactor to man.

Excessive zeal, even with the best of motives, leads to the dominance of imagination over reason, and this realization leads Johnson to agree with Swift about the danger of visionary schemes, in which "some particular train of ideas fixes the attention—and fictions begin to operate as realities."[69] Scientific theories and projects provide frequent examples of this problem, but Johnson, like Swift, immediately extends the notion to political reforms and utopian schemes. In response to the astronomer's confession, the prince, Rasselas, acknowledges an even more dangerous fantasy:

I have frequently endeavoured to image the possibility of a perfect government, by which all wrong should be restrained, all vice reformed, and all the subjects preserved in tranquility and innocence. This thought produced innumerable schemes of reformation and dictated many useful regulations and salutory edicts . . . and I start when I think with how little anguish I once supposed the death of my father and my brothers.[70]

In spite of his sympathy for projectors, then, Johnson advocated no more than incremental change in material and political circumstances and no change whatsoever in religion. To seek more was to do violence to human nature and to risk both personal and social disaster.

When all is said and done, men who commit themselves to scientific studies simply show that they have missed the proper ordering of priorities. The mad astronomer, returned to sanity by his social intercourse with Rasselas and his friends, leaves them with this wisdom about the choice of life:

". . . of the various conditions which the world spreads before you, which you shall prefer," said the sage, "I am not able to instruct you. I can only tell you that I have chosen wrong. . . ."[71]

This passage was written by Johnson in a period of severe depression immediately after the death of his mother; and in brighter moods, as we have seen, Johnson could be much more open to the potential utility of scientific knowledge. His pessimistic vision was no less real, however; it was far more powerfully stated and profound, and it was of a piece with an important tradition of High Church thought.

SUMMARY

A detailed look at Tory–High Church attitudes toward science in the eighteenth century indicates substantial ambivalence and even antagonism to scientific pursuits and, even more, to extensions of scientific modes of argumentation into moral and religious domains. It does not indicate that these sentiments were particularly anti-Newtonian, however, nor that they usually involved an ignorant "carping and obscurantist opposition to everything new and modern." It was sometimes quite the opposite: antagonism to science was linked with a belief in both material and intellectual progress. And identifiably Newtonian science escaped the most intense odium, which was reserved for Cartesian, Hermetic, and Epicurean natural philosophies, all of which shared unacceptable emphases on the discovery of final or ultimate causes. To the extent that High Churchmen accepted the doctrine of secondary causes, they were in accord with standard latitudinarian arguments against pre-Newtonian scientific systems.

High Churchmen tended to be skeptical, though, of the metaphysical and moral claims made on behalf of even Newtonian science. And at the hands of men like John Arbuthnot, their philosophical critiques of eighteenth-century science and scientized morality often transcended the sophistication and cogency of the best rationalizations that could be launched by the latitudinarian and Newtonian opposition.

Like the near contemporary French radical and deist Rousseau, the politically and religiously conservative Tory High Churchmen felt that "our morals have been corrupted in proportion to the advancement of our sciences . . . to perfection."[72] But they shared only some of Rousseau's reasoning. In particular, they tended to be sympathetic to a large part of the utilitarian emphasis on applied science, which Rousseau had thought particularly detestable. They sometimes showed opposition to the use of scientific knowledge for the production of increasingly destructive military technologies; but in general, though they placed very low priority on cosmological speculation and pure science, chiding all who sought trivial knowledge of little benefit to mankind, they did speak on behalf of the acquisition of knowledge for the amelioration of the human condition. Even Swift, who expected almost nothing useful to come out of the schemes of the projectors, nonetheless spoke through the Brobdingnagian king to assert that he who could make two blades of grass grow where before there had been one would deserve to be known as the universal benefactor of mankind.

They were completely in agreement with Rousseau, however, in feeling that an overemphasis on scientific concerns led to an intellectual pride and arrogance incompatible with true Christian humility and with the proper subordination of individual belief to both religious and political authority. Rousseau said the scientific intellectuals "smile disdainfully at the old-fashioned words of fatherland and religion, and devote their talents to destroying and debasing all that is sacred among men."[73] This phrase summarizes a good part of the Tory–High Church opposition to the scientific mentality. Most High Churchmen felt that scientific vanity and arrogance were most destructively manifested in the presumption that natural theology was a necessary foundation or adjunct to revealed Christianity. Though Arbuthnot wrote a brief piece of natural theology as a young man, even he most often took the pervasive High Church position that had been set in the 1670s by men like John Standish, who vilified "those false apostles" who "would supplant Christianity with natural theology, making Reason, Reason, Reason, their one holy Trinity."[74] Neither Rousseau nor most of the High Churchmen felt the scientists were intentionally undermining religious belief; but they did feel that, by acknowledging that religious doctrine was open to debate and criticism and subject to arguments that depended on human rather than divine wisdom, they unintentionally cast doubt on Christian doctrines.

In summation, High Church Anglicanism was generally less supportive of science than latitudinarianism: it both robbed men of one important religious motive for doing science—the motive associated with natural theology—and it sought to limit the approved motives for doing science to purely utilitarian ones. In this sense Margaret Jacob is essentially correct in linking Newtonian science with latitudinarian attitudes in England. When High Churchmen did undertake scientific work, however, they were perhaps even more consistently methodological Newtonians than the latitudinarians. There were, for example, no High Church supporters of Descartes, as was Thomas Burnet among the latitudinarians. Though High Churchmen used Newton's works more selectively and emphasized the limitations to scientific knowledge implied by Newton's focus on secondary causes, to the extent that we seek to explain the dominance of Newtonian science over Hermetic, Cartesian, or other mechanical alternatives, High Church impulses were at one with those of the latitudinarians. On this score Margaret Jacob's suggestion that latitudinarian attitudes alone account for the acceptance of Newtonianism are misleading in the extreme.

NOTES

1. Margaret Jacob, *The Newtonians and the English Revolution: 1689–1720* (Ithaca: Cornell University Press, 1976), p. 51.

2. Cited by Stephen O. Mitchell in "Samuel Johnson and the New Philosophy" (Ph.D. Diss., Indiana University, 1961), p. 59, n. 77. Macaulay's legacy can be beautifully observed in Joseph M. Levine, *Dr. Woodward's Shield: History, Science, and Satire in Augustan England* (Berkeley, Los Angeles, London: University of California Press, 1977). Levine views virtually all of the satirical responses to scientific writings as works of proponents of the "ancients" who were engaged in a "mindless appeal to authority" (p. 10) against the "moderns."

3. G. A. Aitken, *The Life and Works of John Arbuthnot* (London, 1892), p. 9.

4. Lester M. Beattie, *John Arbuthnot: Mathematician and Satirist* (Cambridge, 1935), p. 201.

5. Ibid., p. 198.

6. Ibid.

7. Ibid., p. 203.

8. Quoted in Beattie, *John Arbuthnot*, p. 40.

9. Aitken, *Life and Works of Arbuthnot*, p. 294.

10. See C. Kerby Miller's introduction to *The Memoirs of Martinus Scriblerus* (New Haven, 1950), p. 14 for a discussion of the circumstances surrounding the proposed collaboration.

11. Swift's *Correspondence* II, pp. 162–163, quoted in *Memoirs of Martinus Scriblerus*, p. 57.

12. *Memoirs of Martinus Scriblerus*, p. 168.

13. Ibid., p. 167.

14. Ibid., p. 146.

15. Ibid., pp. 146–147.

16. Ibid., p. 149.

17. Beattie, *John Arbuthnot*, p. 399.

18. *Memoirs of Martinus Scriblerus*, p. 285, n. 13.

19. Aitken, *Life and Works of Arbuthnot*, pp. 438–439.

20. Arbuthnot to Swift, 5 November 1726, reprinted in *Gulliver's Travels*, Norton edition, pp. 266–267.

21. Cited in Middleton Murry, *Jonathan Swift* (London, 1954), p. 100.

22. Ibid., p. 95.

23. *The Prose Works of Jonathan Swift, D.D.*, ed. Temple Scott (London, 1919), III:307–308.

24. *The Works of Sir William Temple*, 4 vols. (London, 1814), III:509.

25. Quoted by Earnest Tuveson in "Swift and the World Makers," *Journal of the History of Ideas*, II (1950): 73.

26. "To the Athenian Society," in *The Poems of Jonathan Swift*, ed. Harold Williams (Oxford, 1937), I:21.

27. *Prose Works*, IV:172–173.

28. See "The Scientific Background of Swift's 'Voyage to Laputa,'" *Annals of Science* 2 (1937): 299–334, 405–430.

29. "Swift and Science," *The Historical Journal* 14 (1971): 722.

30. *Gulliver's Travels*, ed. Robert Greenberg (New York: W. W. Norton, 1970), p. 169.

31. See R. Olson, "Science, Scientism, and Anti-Science in Hellenic Athens: A New Whig Interpretation," *History of Science* 16 (1978): 179–199.

32. *Prose Works*, I:48.

33. Ibid., pp. 53–54.

34. Ibid., p. 92.

35. Ibid., pp. 120–121.

36. Ibid., p. 116.

37. Ibid., pp. 118–119.

38. Ibid., p. 118.

39. Ibid., p. 61.

40. Ibid., p. 61.

41. Ibid., p. 62.

42. See Pat Rogers, "Gulliver and the Engineers," *Modern Language Review* 70 (1975): 260–270.

43. Swift, *Gulliver's Travels*, pp. 132–133.

44. Ibid., p. 150.

45. Ibid., pp. 136–137.

46. Ibid., p. 110.

47. Ibid., p. 111.

48. Ibid., p. 18.

49. John Ray, *The Wisdom of God, Manifested in the Works of the Creation*, 5th edition (London, 1709), pp. 65–66.

50. *Gulliver's Travels*, p. 90.

51. Donald Greene, *Samuel Johnson* (New York: Twayne Publishers, 1970), p. 154.

52. Cited ibid., p. 97.

53. Cited ibid., p. 95.

54. *Letters* III, 54, #870, cited in Mitchell, "Samuel Johnson," p. 63.

55. Samuel Johnson, *Lives of the English Poets* (London, 1779).

56. Cited in Richard S. Westfall, *Science and Religion in Seventeenth Century England* (Ann Arbor: University of Michigan Press, 1973), p. 45.

57. Cited in Chester F. Chapin, *The Religious Thought of Samuel Johnson* (Ann Arbor: University of Michigan Press, 1968), p. 64.

58. Cited in Mitchell, "Samuel Johnson," p. 41.

59. Cited in Greene, *Samuel Johnson*, p. 121, emphasis mine.

60. Ibid., p. 132.

61. Ibid., p. 130.

62. Ibid.

63. Ibid., p. 143.

64. *Idler, 17*, emphasis mine.

65. Cited in J. R. Philip, "Samuel Johnson as Anti-scientist," *Notes and Records of the Royal Society of London* 29 (1975): 195.

66. Samuel Johnson, *Rasselas, Poems, and Selected Prose*, ed. Bertrand H. Bronson (New York: Rinehart, 1952), pp. 590–591.

67. Ibid., p. 592.

68. Ibid., p. 593.

69. Ibid., p. 593.

70. Ibid., p. 597.

71. Ibid., p. 603.

72. Jean Jacques Rousseau, *First and Second Discourses*, ed. Roger D. Masters (New York: St. Martins Press, 1964), p. 39.

73. Ibid., p. 50.

74. Jacob, *The Newtonians* (n. 1 above), p. 47.

Designer:	UC Press Staff
Compositor:	Trend Western
Printer:	Braun-Brumfield
Binder:	Braun-Brumfield
Text:	10 Baskerville
Display:	Baskerville